# A MANUAL

OF

# ELECTRO-METALLURGY:

INCLUDING

## THE APPLICATION OF THE ART

TO

## MANUFACTURING PROCESSES.

BY

JAMES NAPIER, F.C.S.

FOURTH AMERICAN,

FROM THE

FOURTH LONDON EDITION, REVISED AND ENLARGED.

---

ILLUSTRATED BY NUMEROUS ENGRAVINGS.

---

PHILADELPHIA:

HENRY CAREY BAIRD,

INDUSTRIAL PUBLISHER,

406 Walnut Street.

# PUBLISHER'S NOTICE.

---

THE fact of this Treatise being reprinted from the plates of BYRNE'S PRACTICAL METAL WORKER'S ASSISTANT will sufficiently account for the numbering of the chapters, pages, and engravings appearing to be somewhat irregular. It may here be as well to state that the present is an exact and faithful reprint of the Fourth London Edition.

PHILADELPHIA, Dec. 2, 1867.

# PREFACE.

THE author of the following Treatise was engaged for several years in the application of Electro-Metallurgy to the purposes of manufacture. His operations were performed with solutions of all the metals, and upon objects of every size and form. They commenced when the art was young—when its practical applications were speculative—its advantages and disadvantages equally unknown; when difficulties of all kinds, such as beset every new art, had to be met, and considered, and overcome.

The course of his daily proceedings threw him into the way of observations, much more extensive and much more diversified, than could possibly have occurred to any amateur of the art. Where large operations in an extensive business were concerned, it was necessary to attend to details that some would have considered trifling, and to overcome obstacles that others might have deemed insurmountable. Under the pressure of these circumstances, all means were employed to procure information. Innumerable electrotype processes were repeated as soon as they were published, and original experiments were made in a variety of forms, and frequently on an extensive scale, with a view to the removal of particular difficulties, or to find the means of accomplishing certain desirable ends.

These proceedings and inquiries afforded numerous results, not only useful in the manufacture in which the author was engaged, but interesting to the man of science. And it is because of their general utility to all persons engaged in the multifarious processes into which the art of Electro-Metallurgy has ramified, that he has been induced to throw them into the form of the following Treatise.

While, however, he can state, that what is collected here is derived from extensive personal experience, he by no means ventures to assume that the work is free from deficiencies. He has too frequently had to deplore the effects of his processes, and to point out the desirableness of others of greater certainty and economy.

Neither can the author presume that this work will be a *standard* on the subject to which it relates. Arts and sciences, like kingdoms and nations, have their several periods of rise, prevalence, and decadence; and nothing can be more unstable than descriptions of an art like Electro-Metallurgy—an art that must fluctuate with the course of experimental discovery, that has rapidly attained a distinguished eminence, and that promises to extend its utility still farther over regions now unthought of. The superb specimens of its products, which were displayed to the admiration of the world, at the GREAT EXHIBITION OF THE INDUSTRY OF ALL NATIONS, prove at once the immense importance of Electro-Metallurgy, and how much may yet be expected from one of the most ingenious of those modern applications of science, which subject the powers of nature to the use and pleasure of civilized man.

GLASGOW, May, 1851.

---

## NOTE TO THE THIRD EDITION.

The rapid sale of two editions of the Manual of Electro-Metallurgy has afforded great gratification to the author, and he has embraced the present opportunity of revisal to include every approved novelty, so as to bring the work down to the present state of knowledge on this most valuable art.

GLASGOW, February, 1857.

# CONTENTS.

# MANUAL

OF

# ELECTRO-METALLURGY.

---

## CHAPTER XXIV.

### HISTORY OF THE ART OF ELECTRO-METALLURGY.

In reviewing the rise and progress of any discovery in the arts and sciences, particularly of one connected with the application of chemistry to manufacturing purposes, there are two circumstances which almost invariably demand especial notice. The first is, that the discovery has been the result of accidental observation—a fact eliminated during investigations undertaken for other purposes—rather than the result of a direct endeavor to make the discovery. The second is, that, after the discovery has been made known, it is found that many previously published experiments exhibited results which bore more or less directly upon the subsequent discovery, and which are consequently sometimes cited to detract from the merit of the discoverer, and the originality and value of his discovery. The following historical sketch will show that these observations directly apply to the discovery of the art of Electro-Metallurgy:—

Volta's Discovery.—At the beginning of the year 1800, Professor Volta invented the Apparatus which has been named after him, the *Voltaic Pile.* As originally constructed by Volta, it consisted of an equal number of round pieces of zinc, silver, and paste-board—the zinc and silver pieces being each about the size of a penny, and those of pasteboard a little smaller; the pasteboard pieces were soaked in a solution of common salt, and then with the metals were piled in the following manner:—zinc, silver, pasteboard; zinc, silver, pasteboard; and so on, in the same order, till all the pieces, amounting to upwards of a hundred, were piled upon each other, the uppermost plate being of silver, and, as already stated, the undermost of zinc; these exterior plates, to each of which a wire is attached, form the terminals or poles of the pile. Fig. 540 shows the construction of the Pile.

Fig. 540

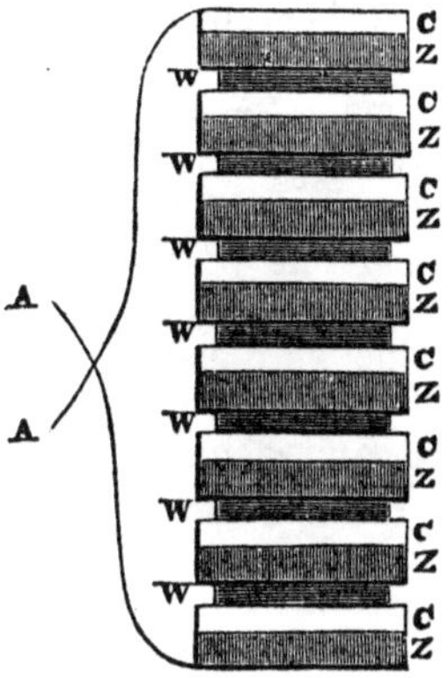

C—Silver Plate.

Z—Zinc Plate.

W—Pasteboard between.

AA—The Wires in connection with the Terminal Plates.

By this instrument all the phenomena of an ordinary battery can be produced.

Chemical Decompositions by the Pile.—This discovery placed in the hands of the philosopher an instrument by which he could make such investigations as had never previously been conceived to be possible. Nicholson, for example, effected the decomposition of water and of several metallic salts: and observed, as a general rule, that in the decomposition of the latter the metal of the salt was reduced upon the zinc terminal of the pile.

First Battery.—Cruikshanks, of Woolwich, with a view to facilitate the construction of the pile, employed square plates of copper and zinc, soldered together two and two; these were cemented, by means of pitch, into a wooden trough, at the distance of about a quarter of an inch from each other, and so arranged that the zinc plates all faced one end of the trough, and the copper plates the other end. The spaces or cells between every pair of plates were filled with a solution of common salt, or a mixture of acid and water, which produced the same effect as the moist cards in the pile. The trough thus charged with its metals and solution acted the same part as the Voltaic pile. This was the first of those instruments now so well known as the "*galvanic battery.*"

Decomposition by the Battery, and its Application.—Cruikshanks attached a silver wire to each terminal of his battery, and the other ends of these wires he placed in a glass tube. When this tube was filled with a solution of acetate of lead, and the electric current was allowed to pass through it for some time, metallic lead was found deposited upon the wire attached to the zinc terminal of the battery. Solutions of sulphate of copper, nitrate of silver, and several other salts, were tried with similar results. The metals, as Cruikshanks expressed it, were "*revived*," and that so completely, as to suggest to him the application of the battery to the analysis of minerals. While Cruikshanks, Nicholson, and several other gentlemen in this country were making investigations and applications of voltaic electricity, upon the Continent, Brugnatelli, Fourcroy, Vauquelin, and Thenard were making similar investigations, and obtaining similar results.*

Deposition of Metals upon others.—Brugnatelli, in his Annals of Chemistry, gives a long list of experiments on the decomposition of salts by the pile. He observed the transfer of the elements of a decomposed compound from one pole to another—that silver, when deposited upon platinum, preserved all its metallic brightness—and that, when copper or zinc were used in connection with the silver terminal, or positive pole, of the pile for decomposing salts, these metals were dissolved, and deposited upon the negative pole. The researches of Fourcroy, Vauquelin, and Thénard gave the same results.

Gilding.—In 1805, Brugnatelli, in a letter to Van Mons, mentions, among other scientific facts, that "he had gilt in a complete manner two large silver medals, by bringing them, by means of a steel wire, into communication with the negative pole of a voltaic pile, and keeping them one after the other immersed in ammoniuret of gold newly made and well saturated."†

Early Opinions concerning Electro-Decomposition.—The above few instances are selected from a host of a similar kind upon electro-decomposition, to show that the fact of the deposition of metals by an electric current was familiar to philosophers at this early stage of the history of galvanism; that nevertheless, the phenomenon was never thought of further than as a curious action of electricity when passing through a solution containing metals; and that although these effects were produced again and again, it was only to prove and enforce certain speculative views respecting the electric fluid. As for example, Brugnatelli had formed an idea that the electric fluid had some relations to an acid which he called the *electric acid*, and he therefore viewed the decomposition of solutions, and the obtaining of the metal, which he termed an *electrate*, as the result of the combination of this electric acid with the metal of the solution. In one of his memoirs upon this subject, he says—"Gold and platinum are not sensibly altered by the electric matter which passes through them, though it often

---

* Wilkinson, Elements of Galvanism, vol. ii. 1804.
† Phil. Magazine, 1805.

happens that the electric current deposits on gold a stratum of zinc copper, mercury, or silver, according to whichever of these metallic bodies it traverses."* In the same paper it is several times stated that gold and platina do not seem sensibly affected by the *electric acid.* And, when he communicated the above experiment of gilding the two medals, his object was to show that he had now found that the electric acid had also the power of acting upon gold; and the publication of these results and observations excited no other idea in the minds of philosophers of that period than that they were mere scientific curiosities. The editor of the *Philosophical Magazine* appended the following note to the extract already quoted:—"The result here detailed reminds me of one, somewhat similar, which took place during some experiments performed some years ago in the Askesian Rooms. Some gold leaf was put loose upon a new piece of copper coin, which was then brought into the circuit of the pile. A part of the gold was inflamed, and other portions adhered to the surface of the copper, as completely as if they had been attached by any common gilding process."†

HOW THESE RESULTS AFFECT THE DISCOVERY.—We have been particular in thus noticing the observations of the first pioneers in electro-chemistry, because these and similar facts of later date have been brought prominently forward by writers upon electro-metallurgy, with the apparent intention to detract from the merit due to the discoverers of the new art; founding their objection on the ground that the *principle* upon which the discovery is founded is not new. "Electro-metallurgy," says Mr. Smee, "may be said to have its origin in the discovery of the constant battery by Professor Daniell, for in that instrument the copper is continually reduced upon the negative plate." And again, when speaking of Daniells' battery, he says—"Mr. M. De la Rue experimented on its properties, and found the copper plate also covered with a coating of metallic copper, which is continually being deposited; and so perfect is the sheet of copper thus formed, that being stripped off, it has the counterparts of every scratch of the plate on which it is deposited."‡

Doubtless these experiments border very closely upon the discovery; but yet they have no more claim to serve as dates to its origin than those we have been referring to. But if it be necessary that an originating experiment must have a resemblance to that which it suggests—such as Daniell's battery; and the single cell of electro-metallurgy—why omit to refer to Dr. Wollaston's earlier experiments of 1801? He says—"If a piece of silver, in connection with a more positive metal, be put into a solution of copper, the silver is coated over with the copper, which coating will stand the operation of burnishing."§ But in our opinion none

* Brugnatelli, Annals of Chemistry, vol. xviii., and Wilkinson, vol. ii.
† Phil. Mag., 1805.
‡ Smee, Elements of Electro-Metallurgy, 2d Edition, 1843.
§ Philosophical Transactions, 1801.

of these results originated electro-metallurgy: the discovery of that art, although it is an application of such results as we have described, was as original on the part of the discoverers, and as unconnected with these results at the time it was made, as it would have been had the earlier observations never been published. The discovery seems to have been deduced from results which the discoverers had obtained in their own experiments, not even while searching for such a discovery but during investigations instituted for other purposes.

USE OF OBSERVED FACTS.—It must not be supposed that we depreciate the value of the published facts upon the decomposition of salts, nor that we overlook their relation to the discovery which followed; for the multiplication of facts, and the improvement of instruments for experimenting, enlarge our knowledge of the principles to be investigated or applied; they facilitate inquiry, and increase the number of observers. The circumstances connected with the discovery of ELECTRO-METALLURGY—of the application of the decomposing force of an electro current passing through a solution, will illustrate these observations.

SPENCER'S FIRST EXPERIMENTS.—Mr. Thomas Spencer, of Liverpool, states that, in 1837, while experimenting with a modification of a Daniell's battery, he used a penny piece instead of a plain piece of copper, as a pole. Copper was deposited from the solution upon it, and on removing the wire which attached the penny to the zinc plate he also pulled off a portion of the deposited copper, which he found to be an exact counterpart or mould of a part of the head and letters of the coin *as smooth and sharp as the original.* But this did not suggest to him any useful application, until sometime after he dropped, accidentally, a little varnish upon a slip of copper which he was about to use in the same way as he had used the penny piece. On finding that no deposit of copper took place on the parts where the varnish had dropped, he *then* conceived the idea of applying this principle to the arts, by coating a piece of copper with varnish or wax, and cutting a design through the wax or varnish, leaving the copper bare, and then depositing upon these parts, so that upon removing the varnish the design would be left in relief.

JACOBI'S EXPERIMENTS.—While Mr. Spencer was following up these ideas, the following paragraph appeared in the *Athenæum* for 4th May, 1839:

"*Galvanic Engraving in Relief.*—While M. Daguerre and Mr. Fox Talbot have been dipping their pencils in the solar spectrum, and astonishing us with their inventions, it appears that Professor Jacobi, at St. Petersburgh, has also made a discovery which promises to be of little less importance to the arts. He has found a method—if we understand our informant rightly—of converting any line, however fine, engraved on copper, into a relief, by galvanic process. The Emperor of Russia has placed at the Professor's disposal, funds to enable him to perfect his discovery."

In consequence of this announcement, Mr. Spencer, on the 8th of May, 1839, gave notice to the Liverpool Polytechnic Institution, that he should make a communication to them of his process for effecting results similar to those of Professor Jacobi. But Mr Spencer appears to have changed his design of reading it to the above Institution, in order to have it read at the meeting of the British Association, which was to take place a short time after.

JORDAN'S EXPERIMENTS.—Meanwhile the announcement of the *Athenæum* was quoted in the *London Mechanics' Magazine* for May 11th, 1839, which brought forth a letter from Mr. C. J. Jordan, a book-printer, dated 22d May, 1839, and published on the 8th June of the same year in the *London Mechanics' Magazine*. In this letter Mr. Jordan describes his experiments upon the same subject, detailing the method of procuring electrotypes, and offering hints for their application which have since been acted upon with considerable success. The following is a copy of Mr. Jordan's letter, which was, no doubt, the first published description of the art in this country:

"*Engraving by Galvanism.*

"Sir:—Observing in the last page of a recent Number of your Magazine, a notice extracted from the *Athenæum*, relative to a discovery of Professor Jacobi, its perusal occasioned the recollection of some experiments performed about the commencement of last summer, with the view of obtaining impressions from engraved copper plates, by the aid of galvanism, which led me to infer some analogy in principle with those of the Russian Professor, and may probably give me the right to claim priority in its discovery and application. These experiments were abandoned from the want of that most important element in pursuits of this nature—time; the writer's share of the said element being occupied in a manner more imperative than pleasing. I regret, however, not having made it the subject of an earlier communication, as this would have placed my pretensions beyond doubt; but, inasmuch as the notice alluded to is given from memory, and is undescriptive, while I may be enabled to exhibit the *modus operandi*, my assertion may be at least partially substantiated.

"It is well known to experimentalists on the chemical action of voltaic electricity, that solutions of several metallic salts are decomposed by its agency, and the metal procured in a free state. Such results are very conspicuous with copper salts, which metal may be obtained from its sulphate (blue vitriol), by simply immersing the poles of a galvanic battery in its solution, the positive wire becoming gradually coated with copper. This phenomenon of metallic reduction is an essential feature in the action of sustaining batteries, the effect, in this case, taking place on more extensive surfaces. But the form of voltaic apparatus which exhibits this result in the most interesting manner, and relates more immediately to the subject of the present communication, may be thus described:—It consists of a glass tube, closed at one extremity with a plug of plaster of Paris, and nearly filled with a solution of sulphate of copper; this tube and its contents are immersed in a solution of common salt. A plate of copper is placed in the first solution, and is connected, by means of a wire and solder, with a zinc plate which dips into the latter. A slow electric action is thus established through the pores of the plaster, which it is not necessary to mention here—the result of which is the precipitation of minutely crystallized copper on the plate of that metal, in a state of greater or less malleability according to the slowness or rapidity with which it is deposited. In some experiments of this nature, on removing the copper thus formed, I remarked that the surface in contact with the plate equalled the latter in smoothness and polish, and mentioned this fact to some individuals of my acquaintance. It occurred to me, therefore, that if the surface of the plate was engraved, an impression might be obtained. This was found to be the case; for, on detaching the precipitated metal, the most delicate and superficial markings, from the fine particles of powder used in polishing, to the deeper touches of a needle or a graver, exhibited their correspondent impressions in relief, with great fidelity. It is, therefore, evident that this principle will admit of improvement, and that casts and moulds may be obtained from any form of copper.

"This rendered it probable that impressions may be obtained from those other metals having an electro-negative relation to the zinc plate of the battery. With this view, a common printing-type was substituted for the copper plate, and treated in the same manner. This also was successful; the reduced copper coated that portion of the type immersed in the solution. This, when removed, was found to be a perfect matrix, and might be employed for the purpose of casting, where time is not an object.

It appears, therefore, that this discovery may be turned to some practical account. It may be taken advantage of in procuring casts from various metals, as above alluded to; for instance, a copper die may be formed from a cast of a coin or medal, in silver, type metal, or lead, etc., which may be employed in striking impressions in soft metals. Casts may probably be obtained from a plaster surface surrounding a plate of copper; tubes or any small vessels may also be made by precipitating the metal around a wire, or any kind of surface, to form the interior, which may be removed mechanically, by the aid of an acid solvent, or by heat.

"C. J. JORDAN."

"*To the Editor of the London Mechanics' Magazine.*"

Clear and perspicuous as this letter is, it did not attract the slightest notice. And a few weeks after, we find that its existence was forgotten even by the editor of the magazine in which it appeared.

SPENCER'S FIRST PRINTED PAPER UPON ELECTROTYPE.—Mr. Spencer's communication, referred to above, was, in consequence of some misunderstanding, not read at the meeting of the British Association, but it was immediately afterwards read before the Polytechnic Institution of Liverpool, at their meeting on the 13th September, 1839, which was upwards of three months after the publication of Mr. Jordan's letter in the *London Mechanics' Magazine.* Mr. Spencer's paper was accompanied with specimens both of electrotypes and of printing from electrotypes. The publication of this paper acted like an electric shock upon society, and men both of science and art became active competitors in this new field of application; the one class anxious to bear away the honors arising from some important improvement; the other, the profits which might follow some novel application of the process to their own or some other branch of manufacture. Indeed, thousands of all classes and ages, who had never previously given science a passing thought, became fascinated with the new art, and—the process being simple and easy to perform—the amateurs soon became excellent electrotypists. With these combined efforts, it need not be wondered at that in a very short time improvements of great scientific interest were pointed out, and applications of the greatest importance to the arts and manufactures of this country were introduced. In consequence, some of our old and standard manufactures, as we shall subsequently have occasion to notice at some length, have already been revolutionized.

HISTORICAL ANOMALY.—During a period of nearly five years—while the country was passing through an electrotyping mania—Mr. Spencer held the undivided honor of being the first to apply the deposition of metals to practical purposes in this country, but early in 1844, Mr. Henry Dircks, in a letter to the *London Mechanics' Magazine*, revived Mr. Jordan's letter, and told us that he was aware of its existence from the time of its first publication. We cannot eulogise either the policy, or the love of scientific truth, which induced Mr. Dircks to remain silent so long, and see the claims of Mr. Jordan set aside by one whom he considered to be a mere pretender to the merit of the discovery. Nor, after a careful and impartial examination of all the details published on the subject, can we agree to his condemnation of Mr. Spencer's prior claims; as he, Mr. Spencer, upon the 8th of May, as already noticed, stated to a public meeting of the Polytechnic Society of Liverpool, that he had made a similar discovery previous to any knowledge of either what Jordan or Jacobi had done, and which was a publication as much as if printed in the *Times* or *Athenæum*, and especially when followed by a detailed description of the discovery.

It is to be regretted that Mr. Jordan's diffidence, which in this case was far from being commendable, prevented his setting the public right upon this important matter. As a consequence, he must now be content with a much smaller share of the honor of the discovery than he might have enjoyed.

On reviewing the circumstances of this discovery, it strikes us as being a remarkable instance of the unity of intellectual perception in reference to the general principles of Nature and their applications; for we believe that Professor Jacobi, Mr. Spencer, and Mr. Jordan, viewed the subject of electro depositions in the same light, and about the same time; and each, according to their several abilities, presented to the public the same discovery, independent of the other, excepting the announcement made by one having hastened the publication of the observations of the others.

The following is Mr. Spencer's original paper on electro-metallurgy, which we give at length, trusting that its importance in connection with the history of the art, and the lucid description of its practice, will serve as a sufficient apology for not abridging it:

"*On Working in Metal by Voltaic Electricity, reprinted from the Paper Published by the Liverpool Polytechnic Society, and read at the Meeting of September the* 12*th*, 1839, *Notice being given May the* 8*th: Henry Booth, Esq., President, in the Chair.*

"In the paper that I have the honor to lay before the society, I do not profess to have brought forward a perfect invention. My only object is to point out a means by which, I hope, practical men may be enabled to apply a great and universal principle of Nature to the useful and ornamental purposes of life. In this I may be considered sanguine—an error, I am aware, too often fallen into by those who, like myself, imagine they have discovered a useful application of an important principle; but however this may fall out, I shall lay an account of its results, with specimens, successful and unsuccessful, before the members and the public—previously stating, however, that all my first experiments were made on a small scale—a method of procedure attended with many advantages to the experimentalist himself, but having its disadvantage when laid before the public. In this first respect, perhaps, the chemical experimenter has an advantage over the mechanical one, as the success of his experiment, when tried on a small scale, doubly guarantees it if conducted on a still larger scale: with mechanical results I believe in most instances it is the reverse. But when the chemist produces his microscopic proofs, the public are generally slow to believe that such minute appearances should warrant him in coming to any general conclusion.

"In the latter part of September, 1837, I was induced to make some electro-chemical experiments, with single pairs of plates, consisting of small pieces of zinc and equal-sized pieces of copper, connected together with wires of the latter metal. It was intended that the action should be slow: the fluids in which the metallic electrodes were immersed were in consequence separated by thin discs of plaster of Paris. In one cell thus formed was placed sulphate of copper in solution—in the other, a weak solution of common salt. I need scarcely add that the copper electrode was placed in the cupreous solution, the other being in that of the salt.

I mention these experiments briefly—not because they are *directly* connected with what I shall have to lay before the society, but because, by a portion of their results, I was induced to come to the conclusions I have done in the following paper. I was desirous that no action should take place on the wires by which the electrodes were held together; and to attain this object I varnished them with sealing-wax varnish: but, in one instance, I dropt a portion on the copper electrode that was attached. I thought nothing of this circumstance at the moment, but put the experiment in action.

"This operation was conducted in a glass vessel; I had consequently an opportunity of occasionally examining its progress from the exterior. After the lapse of a few days, metallic crystals had covered the copper electrode—*with the exception of that portion* which had been spotted with the drops of varnish. I at once saw that I had it in my power to guide the metallic deposition in any shape or form I chose, by a corresponding application of varnish or other non-metallic substance.

"I had been aware of what every one who uses a sustaining galvanic battery with sulphate of copper in solution must know—that the copper plates acquire a coating of copper from the action of the battery; but I had never thought of applying it to a useful purpose, except to multiply the plates of a species of battery, which I did in 1836. My present attempt was with a piece of thin copper plate, having about four inches of superfices, with an equal sized piece of zinc, connected as before by a piece of copper wire. I gave the copper a coating of soft cement, consisting of beeswax, resin, and a red earth. It was compounded in the manner recommended by Dr. Faraday, in his work on Chemical Manipulation, but with a larger proportion of wax. The plate received its coating while hot. When it was cold, I scratched the initials of my name rudely on the plate, taking special care that the cement was quite removed from the scratches, that the copper might be thoroughly exposed. This was put in action in a cylindrical glass vessel, about half filled with a saturated solution of sulphate of copper. I then took a common gas glass, similar to that used to envelope an argand burner, and filled one end of it with plaster of Paris, to the depth of three-quarters of an inch. Into this I put water, adding a few crystals of sulphate of soda to excite action, the plaster of Paris acting as a partition to separate the fluids, but, at the same time, being sufficiently porous to allow the electro-chemical action to permeate its substance.

"I now bent the wire in such a manner that the zinc end of the arrangement should be in the saline solution, while the copper end, when in its place, should be in the cupreous solution. The gas glass, with the wire, was then placed in the vessel containing the sulphate of copper.

"It was then suffered to remain at rest, when in a few hours I perceived that action had commenced, and that the portion of the copper rendered bare by the scratches had become gradually coated with pure bright deposited metal, whilst all the surrounding portions were not at all acted on. I now saw my former observations realized; but whether the deposition so formed would retain its hold on the plate, and whether it would be of sufficient solidity or strength to bear working if applied to a useful purpose, became questions which I now determined to solve by experiment. It also became a question—should I be successful in these two points—whether I should be able to produce lines sufficiently in relief to print from. This latter appeared to depend entirely on the nature of the cement or etching-ground I might use.

"This I endeavored to solve at once; and, I may state, it appeared at the time to be the main difficulty, as my impression then was, that little less than one-eighth of an inch of relief would be requisite to print from.

"I now procured a piece of copper, and gave it a coating of a modification of the cement I have already mentioned, and having covered it to about one-eighth of an inch in thickness, I took a steel point and endeavored to draw lines in the form of net-work, that should entirely penetrate the cement, and leave the surface of the copper exposed. But in this I experienced much difficulty, from the thickness I deemed it necessary to use; more especially when I came to draw the cross lines of the net-work. The cement being soft, the lines were pushed as it were into each other, and when it was made of harder texture, the intervening squares of the net-work chipped off the surface of the metallic plate. However those that remained perfect I put in action as before.

"In the progress of this experiment I discovered that the solidity of the metallic deposition depended entirely on the weakness or intensity of the electro-chemical action, which I knew I had in my power to regulate at pleasure, by the thickness of the intervening wall of plaster of Paris, and by the coarseness or fineness of the material. I made three similar experiments, altering the texture and thickness of the plaster each time, by which I ascertained that if the partitions were *thin* and *coarse*, the metallic depositions proceeded with great *rapidity*, but the crystals were friable and easily separated; on the other hand, if I made them thicker and of a little finer material, the action was slower, but the metallic deposition was as solid and ductile as copper formed by the usual methods —indeed, when the action was exceedingly slow, I have had a metallic deposition apparently much harder than common sheet copper, but more brittle.

"There was one most important and, to me, discouraging circum stance attending these experiments, which was, that when I heated the plates to get off the covering of cement, the meshes of copper net-work occasionally *came off with it.* I at one time imagined this difficulty inseparable, as it appeared that I had cleared the cement entirely from the surface of the copper that I meant to have exposed; and I concluded that there must be difference in the mole cular arrangement of copper prepared by heat and that prepared by voltaic action, which prevented their chemical combination. However, I determined, should this prove so, to turn it to account in another manner, which I shall relate in the second portion of the paper.

"I now occupied myself for a considerable period in making experiments on this latter section of the subject.

"In one of them I found, on examination, that a portion of the copper deposition, which I had been forming on the surface of a coin, adhered so strongly that I was quite unable to get it off—indeed, a chemical combination had apparently taken place. This was only on one or two spots on the prominent parts of the coin. I immediately recollected that, on the day I put the experiment in action, I had been using nitric acid for another purpose, on the table I was operating on, and that in all probability the coin might have been laid down where a few drops of the acid had accidentally fallen. Bearing this in view, I took a piece of copper, coated it with cement, made a few scratches on its surface until the copper appeared, and immersed it for a short time in dilute nitric acid, until I perceived, by an elimination of nitrous gas, that the exposed portions were acted upon sufficiently to be slightly corroded. I washed the copper in water, and put it in action as before described. In forty-eight hours I examined it, and found the lines were entirely filled with copper, I applied heat, and then spirits of turpentine, to get off the cement, and, to my satisfaction, I found that the voltaic copper had completely combined itself with the sheet on which it was deposited.

"I then gave a plate a coating of cement to a considerable thickness, and sent it to an engraver; but when it was returned I found the lines were cleared out so as to be wedge-shaped, or somewhat in the form of a V, leaving a hair line of the copper exposed at the bottom, and a broad space near the surface; and where the turn of the letters took place, the top edges of the lines were galled and rendered rugged by the action of the graver. This, of course, was an important objection, which I have since been able to remedy in some degree by an alteration in the shape of the graver, which should be made of a shape more resembling a narrow parallelogram than those in common use: some engravers have many of their tools so made. I did not put this plate in action, as I saw that the lines, when in relief, would have been broad at the top and narrow at the bottom. I took another plate, gave it a coating of the wax, and had it written on with a mere point. I deposited copper on the lines, and afterwards had it printed from.*

"I now considered part of the difficulties removed: the principal one yet remaining was to find a cement or etching-ground, the texture of which should be capable of being cut to the re-

* This plate was shown to friends, and also specimens of printing from it, in 1838.

quired depth, without raising what is technically termed *a burr*, and, at the same time, of sufficient toughness to adhere to the plate, when reduced to a small isolated point, which would necessarily occur in the operation which wood-engravers term cross-hatching.

"I have since learned, from practical engravers, that much less relief is necessary to print from than I had deemed indispensable, and that, on becoming more familiar with the cutting of the wax-cement, they would be enabled to engrave in it with great facility and precision.

"I tried a number of experiments with different combinations of wax, resins, varnishes, earths, and metallic oxides, all with more or less success. One combination that exceeded all others in its texture was principally composed of beeswax, resin, and white lead. This had nearly every requisite, so that I was enabled to polish the surface of the plate with it until it was nearly as smooth as a plate of glass. With this compound I had two plates, five inches by seven, coated over, and portions of maps cut on the cement, which I had intended should have been printed off. I applied the same process to these as to the others, immersing them into dilute nitric acid before putting them in action; indeed I suffered them to remain about ten minutes in the solution. I then put them into the voltaic arrangement. The action proceeded slowly and perfectly for a few days, when I removed them. I applied heat as usual, to remove the cement, but *all* came away, as in a former instance—the voltaic copper peeling off the plate with the greatest facility. I was much puzzled at this unexpected result; but, on cleaning the plate, I discovered a delicate trace of *lead*, exactly corresponding to the lines drawn on the cement previous to the immersion in the dilute acid. The cause of this failure was at once obvious: the carbonate of lead I had used to compound the etching-ground had been decomposed by the dilute nitric acid and the metallic lead thus reduced had deposited itself on the exposed portions of the copper plates, preventing the voltaic copper from chemically combining with the sheet copper. I was now with regret obliged to give up this compound, and to adopt another, consisting of beeswax, common resin, and a small portion of plaster of Paris. This seems to answer the purpose tolerably, though I have no doubt, by an extended practice, a better may still be obtained by a person practically acquainted with the etching-grounds in use among engravers.

"I now proceed to the second, and I believe the most satisfactory portion of the subject. Although I have placed these experiments last, some of them were made at the same time with the others already described, and some of them before; but, to render the subject more intelligible, I have placed them thus.

"The members of the society will recollect that, on the first evening it met, I read a paper on the 'production of metallic veins in the crust of the earth,' and that among other specimens of cupreous crystallization which I produced on that occasion, I exhibited three coins—one wholly covered with metallic crystals, the other on one side only. It was used under the following circumstances. When about to make the experiment, I had not a slip of copper at hand to form the negative end of my arrangement, and, as a good substitute, I took a penny and fastened it to one end of the wire, and put it, in connection with a piece of zinc, in the apparatus already described.

"Voltaic action took place, and the copper coin became covered with a deposition of copper in a crystalline form. But, when about to make another experiment, and being desirous of using the piece of wire used in the first instance, I pulled it off the coin to which it was attached. In doing this, a piece of the deposited copper came off with it; on examining the under portion of which, I found it contained an exact mould of a part of the head and letters of the coin, as smooth and sharp in every respect as the original on which it was deposited. I was much struck with this at the time; but, on examination, the deposition metal was very brittle. This, and the fact that it would require a metallic nucleus to aggregate on, made me apprehensive that its future usefulness would be materially abridged; but it was reserved for future experiment, and in consequence laid aside for a time, until my attention was recalled to the subject in a subsequent experiment, already detailed, by the drops of varnish on a slip of copper. Finding in that instance that the deposit would take the direction of any non-conducting material, and be, as it were, guided by it, I was induced to give the previous branch of the subject a second trial, because I had, in the first instance, supposed that the deposition would only take place continuously, and not on isolated specks of a metallic surface, as I now found it would; but the principal inducement to investigate the subject was the fact of finding that deposited copper had much more tenacity than I at first imagined.

"Being aware of the apparent natural law which limits metallic deposition by voltaic electricity, excepting in the presence of a metallic body, I perceived that the uses of the process would, in consequence, be extremely limited, except in the multiplication of already engraved plates, as, whatever ornament it might produce, it would only be done by adhering to the condition of a metallic mould.

"I accordingly determined to make an experiment on a very prominent copper medal. It was placed in a voltaic circuit, as already described, and deposited a surface of copper on one of its sides to about the thickness of a shilling. I then proceeded to get the deposition off. In this I experienced some difficulty, but ultimately succeeded. On examination with a lens, every line was as perfect as the coin from which it was taken. I was then induced to use the same piece again, and let it remain a much longer time in action, that I might have a thicker and more substantial mould, in order to test fairly the strength of the metal. It was accordingly put again in action, and let remain until it had acquired a much thicker coating of the metallic deposition; but on attempting to remove it from the medal I found I was unable. It had, apparently, completely adhered to it.

"I had often practised, with some degree of success, a method of preventing the oxidation of polished steel, by slightly heating it until it would melt fine beeswax; it was then wiped, apparently completely off, but the pores or surface of the metal became impregnated with the wax.

"I thought of this method, and applied it to a copper coin.

"I first heated it, applied wax, and then wiped it so completely off, that the sharpness of the coin was not at all interfered with. I proceeded as before, and deposited a thick coating of copper on its surface. Being desirous to take it off, I applied the heat of a spirit-lamp to the back, when a sharp crackling noise took place, and I had the satisfaction of perceiving that the coin was completely loosened. In short, I had a most complete and perfect copper mould of one side of a half-penny.

"I have since taken some impressions from the mould thus taken, and, by adopting the above method with the wax, they are separated with the greatest ease.

"By this experiment it would appear that the wax impregnates the surface of the metal to an inconsiderable depth, and prevents a chemical adhesion from taking place on the two surfaces; and I can only account for the crackling noise, on separation, by supposing it probable that the molecular arrangement of the voltaic metal is different from that subjected to percussion, and this difference causes an unequal degree of expansibility on the application of heat.

"I became now of opinion, that this latter method might be applied to engraving much better than the method described in the first portion of this paper. Having found in a former experiment that copper in a voltaic circuit deposited itself on lead with as much rapidity as on copper, I took a silver coin, and put it between two pieces of clean sheet-lead, and placed them under a common screw-press. From the softness of the lead, I had a complete and sharp mould of both sides of the coin, without sustaining injury. I then took a piece of copper wire, soldered the lead to one end, and the piece of zinc to the other, and put them into the voltaic arrangement I have already described. I did *not*, in this instance, *wax* the mould, as I felt assured that the deposited copper would easily separate from the lead by the application of heat, from the different expansibility of the two metals.

"In this result I was not disappointed. When the heat of a spirit-lamp was applied for a few seconds to the lead, the copper impression came easily off. So complete do I think this latter portion of the subject, that I have no hesitation in asserting that *fac-similies* of any coin or medal, no matter of what size, may be readily taken, and as sharp as the original. To test further the capabilities of this method, I took a piece of lead plate, and stamped some letters of its surface to a depth sufficient to print from, when in relief. I deposited the copper on it, and found it came easily off, the letters being in relief.

"Finding from this experiment that the extreme softness of lead allowed it to be impressed on by type metal, I caused a small portion of ornamental letter-press to be set up in type, and placing it on a planed piece of sheet lead, it was subjected to the action of a screw press.

"After considerable pressure, it was found that a perfectly sharp mould of the whole had been obtained in the lead. A wire was now soldered to it, and it was placed in an apparatus similar in principle, but larger than the one already described. At the end of eight days from this time, copper was deposited to one-eighth of an inch in thickness; it was then removed from the apparatus, and the rough edges of the deposited copper being filed off, it was subjected to heat, when the two metals began to loosen. The separation was completed by inserting a piece of wedge-shaped wood between them.

"I had now the satisfaction of perceiving that I had by these means obtained a most perfect specimen of stereotyping in copper, which had only to be mounted on a wooden block to be ready to print from.

"From the successful issue of this experiment, which was mainly due to the susceptibility of the lead, I was induced to attempt to copy a wood engraving by a similar method, provided the wood would bear the requisite pressure. Knowing that wood engravings are executed on the *end* of the block, I had better hopes of succeeding, the wood being less likely to sustain injury.

"I accordingly procured a small wood block, and placed its engraved surface in contact with a piece of sheet lead made very clean, and subjected it to pressure, as in the former instance. I had now, as before, the gratification of perceiving that a perfect mould of the little block had been obtained, and no injury done to the original. Several wood engravings and copperplates were subjected to similar treatment, and are now in process of being deposited on in the apparatus before me.

"I now come to the third and concluding portion of the experiments on this subject. The object being to deposit a metallic surface on a model of clay, wood, or other *non*-metallic body—as, otherwise, I imagined the application of this principle would be extremely limited. Many experiments were made to attain this result, which I shall not detail, but content myself with describing those which were ultimately most successful.

"I procured two models of an ornament, one made of clay, and the other of plaster of Paris, soaked them for some time in linseed-oil, took them out, and suffered them to dry—first getting the oil clean off the surface. When dry, I gave them a thin coat of mastic varnish. When the varnish was nearly dry, *but not thoroughly so*, I sprinkled some bronze powder on that portion I wished to make a mould of. This powder is principally composed of mercury and sulphur, or it may be chemically termed a sulphuret of mercury. There is a sort that acts much better, in which is a portion of gold. I had, however, a complete metalliferous coating on the surface of the model, by which I was enabled to deposit a surface of copper on it, by the voltaic method I have already described. I have also gilt the surface of a clay model with gold leaf, and have been successful in depositing copper on its surface. There is likewise another, and as I trust it will prove, a simpler method of attaining this object; but as I have not yet sufficiently tested it by experiment, I shall take another opportunity of describing it."

[At the close of the paper, several specimens of coins, medals, and copper plates, some of them in the act of formation by the voltaic process, were exhibited by Mr. Spencer to the Society.]

---

PLUMBAGO AS A COATING.—Shortly after Mr. Spencer's paper was published, several important improvements were introduced, one or two of which we will refer to here, and will give the others when detailing the processes to which the improvements were applied. The first was the use of plumbago or black lead, to give the surface of non-metallic bodies a conducting property. This was the discovery of Mr. Robert Murray, a gentleman of high attainments and unassuming manners, who communicated the process to the members of the Royal Institution orally. The Society of Arts afterwards awarded to Mr. Murray a silver medal, as an expression of their sense of the value of the discovery. Seldom was reward more deserving, or a discovery more important to the purposes to which it was to be applied, for this application at once freed electro-metallurgy from every bond: it was no longer necessary to use either metallic moulds, or moulds having metal reduced upon their surfaces by chemical means—which, according to the processes then known, was both tedious and uncertain, and only applicable to certain substances. Plumbago possessed all the requisite properties: it was convenient, plentiful, and cheap, easily applied, and equally effective for every substance on which the electrotypist desired to obtain a deposit, or which he could wish to cover with metal, either for useful or ornamental purposes.

SEPARATE BATTERY.—The second improvement to be noticed is one that must have followed the original discovery very soon, namely, the application of a *separate battery* for the purposes of deposition. This was suggested by Mr. Mason. Although those instances of deposition of metals which have been referred to in the early history of galvanism were effected by means of separate batteries, namely, by placing the ends of the wires attached to the terminals of the battery in the solution to be decomposed, still the discovery under consideration was made by means of what is termed the *single cell.* It consisted in simply attaching by a wire the article upon which a deposition was to be made to a piece of zinc, and immersing the zinc in diluted acid, and the other article in a solution of the metal to be deposited; the two liquids being separated by a porous partition, or diaphragm, such as moist bladder, or unglazed porcelain. In this case, the whole electricity was expended within the cell, to deposit the metal within or upon the mould. By Mr. Mason's discovery, the electricity generated in the cell could be made to do an equivalent of work in a separate cell as well—making the original arrangement the generating cell or battery to the second cell. In this last cell was also a solution of a metal having in it a sheet of similar metal attached to the copper of the first cell, and the mould to be covered was attached to the zinc of the first cell. Fig. 541 is an illustration of Mason's improvement, which consisted in causing a medal, in the act of being deposited, to serve as part of a battery for the deposition of another medal.

$O^2$, Outer vessel filled with sulphate of copper; $O^1$, another vessel, with P, a porous cell filled with dilute acid, in which is placed Z a zinc plate, which is connected by a wire with a medal *m'* in the second vessel charged with sulphate of copper. The medal, *m*, in the first cell, is connected by a wire to a piece of copper, C, in the second cell: the electricity passes from the zinc Z to *m*, and by the wire to C, then to *m'*, and by the wire back to the zinc Z.

Fig. 541.

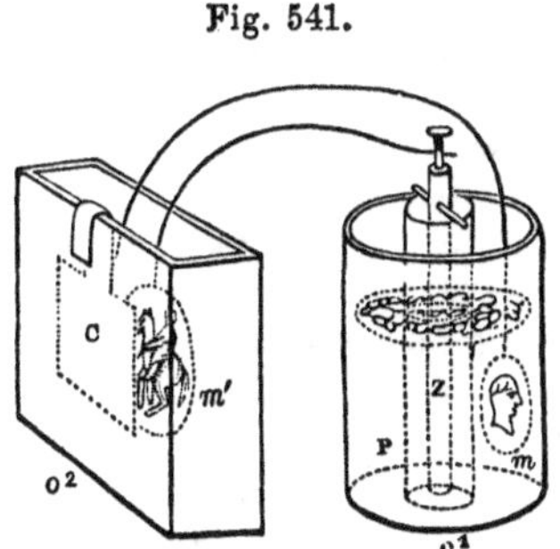


The use of a separate battery, however self-evident, was a valuable addition to the electro-metallurgist for many of his operations; although for some purposes the original single cell is to be preferred.

Laws of Deposition.—As might have been expected in the excitement occasioned by the announcement of a new art, every individual experimenter became so engrossed by his own investigations and their results, as to overlook the labors of others, and at last to lay claim to the honor of originating *all* the discoveries they announced; while the truth is, that nearly all the important facts of electro-metallurgy appear to have occurred almost simultaneously to various experimenters. We shall quote one or two instances of these absorbing claims, as it is important to rectify the errors they contain, because no successful laborer, however humble, in the field of science or art, should be overlooked by his fellow-laborer whose opportunities for research may be much more favorable.

*Extract from Smee's "Electro-Metallurgy."*

"The laws regulating the reduction of all metals in different states were first given in this work as the result of my own discoveries. By these we can throw down gold, silver, platinum, palladium, copper, iron, and almost all other metals, in three states, namely, as a black powder, as a crystalline deposit, or as a flexible plate. These laws appear to me at once to raise the isolated facts known as the electrotype into a science, and to add electro-metallurgy as an auxiliary to the noble arts of this country."

That Mr. Smee discovered the laws referred to we have not the slightest doubt: they were published as laws in his book, and they are commonly quoted as Mr. Smee's; nevertheless, he was not the first who discovered them; the same laws were pointed out by Mr. Spencer, in his original paper just quoted, published eighteen months previously to the appearance of Mr. Smee's work, as will be seen from the following extracts:

*Laws given by Smee.*

"Law I.—The metals are invariably thrown down as a black powder, when the current of electricity is so strong, in relation to the strength of the solution, that hydrogen is evolved from the negative plate of the decomposition cell.

"Law II.—Every metal is thrown down in a crystalline state, when there is no evolution of gas from the negative plate, or no tendency thereto.

"Law III.—Metals are reduced in the reguline state, when the quantity of electricity, in relation to the strength of the solution, is insufficient to cause the production of hydrogen in the negative plate of the decomposition trough, and yet the quantity of electricity very nearly suffices to induce that phenomenon."

*Laws given by Spencer.*

"I discovered that the solidity of the metallic deposition depended entirely on the weakness or intensity of the electro-chemical action, which I knew I had in my power to regulate at pleasure, by the thickness of the intervening wall of plaster of Paris, and by the coarseness or fineness of the material. I made three similar experiments, altering the texture and thickness each time, by which I ascertained, that if the partitions were *thin* and *coarse*, the metallic deposition proceeded with great *rapidity*, but the crystals were friable and easily separated; on the other hand, if I made them thicker, and of a little finer material, the action was slower, but the metallic deposition was as solid and ductile as copper formed by the usual methods. Indeed, when the action was exceedingly slow, I have had a metallic deposition much harder than common sheet copper, but more brittle."

The identity of these deductions or laws requires no comment; and, comparing the circumstances of the one having nothing but the rude apparatus of a new-born art suggested by himself, to that of the other, enjoying the advantage of eighteen months improvements, Mr. Spencer is astonishingly correct, and his name should be identified with the discovery of these laws. The claim of originality involved in the inference drawn by Mr. Smee, though formidable at first sight, is nevertheless, without foundation. Mr. Smee says: "These laws appear to me at once to raise the isolated facts known as the electrotype into a science, *and to add electro-metallurgy as an auxiliary to the noble arts of this country.*" Unfortunately for the validity of Mr. Smee's claim, patents were taken out long previous, both in England and France, for the application of the electro-depositions to the arts. And Messrs. Elkington's patent for silvering and gilding by this process—a patent which has not yet been superseded—was not only published in full detail, but was in extensive operation months before the publication of Mr. Smee's book. Nevertheless, the publication of Mr. Smee's book did good service to the art of electrotyping; and the invention of his battery has so identified his name with the science, that it will go down to posterity as that of an active and successful laborer in the field of electricity: to Mr. Smee we owe, moreover, the very appropriate name for the art, *Electro-metallurgy.*

WORKS PUBLISHED ON ELECTRO-METALLURGY.—Besides many interesting papers in journals and magazines, several separate works were published on this art. Some months previously to the publication of Mr. Smee's work, Mr. Spencer had given the world "*Instructions for the multiplication of works of art in metal by voltaic electricity;*" and shortly after Mr. Smee's work appeared, we had, in rapid succession, Walker's *Electrotype Manipulation,* Sturgeon's *Art of Electrotyping,* Shaw's *Manual of Electro-metallurgy,* etc., all showing much practical knowledge of the subject. Walker's *Manipulation,* from its practical nature and its concise form, became the favorite of the amateur, and did more to popularize the art than all the others put together; and although little pretensions were made to originality, the author will not fail to have an honorable remembrance in the history of the art.

PATENTS TAKEN OUT FOR ELECTRO-METALLURGY.—The application of the art to useful purposes was so self-evident and so eagerly sought after, that no less than ten patents were taken out for useful applications, between the discovery of the art and the close of 1841; and not a year has passed since, without adding patents for certain improvements, and applications of Electro-metallurgy to some particular branch of manufacture, several of which will be noticed in their proper places, as many of them, although based upon right principles, were commercially speaking, entire failures.

# CHAPTER XXV.

## DESCRIPTION OF GALVANIC BATTERIES, AND THEIR RESPECTIVE PECULIARITIES.

Nomenclature.—The terms that are employed to denote the various parts of a galvanic battery, and of other electrotype arrangements, frequently puzzle the student, and lead him into difficulties. Before we proceed to describe the various forms of the battery, we shall, for this reason, give a preliminary account of the nomencla ture of galvanism.

The two extremities of a battery have long been called *Poles*—one of them the *Positive*, and the other the *Negative*, Pole. But objections have been taken to the use of the terms *negative*, *positive*, and *pole*, on the ground that such terms do not convey a correct idea of the circumstances or of the effects produced. Before connecting the two metals or extremities of a battery, there is no electricity evolved, nor is there any electrical tension on any part of the arrangement; and when the connection is formed the electricity simply makes a circuit, it is therefore supposed that no particular portion of that circuit can be said to be either negative or positive to another portion.

Proposed Terms.—Various terms have been suggested as substitutes for negative and positive, and also for pole. Dr. Faraday has proposed the following: for pole, he substitutes *electrode*, which signifies *a way;* for the negative pole, *cathode*, signifying *downwards;* and for the positive pole, *anode* or *upwards*. To understand these terms properly, we must suppose a battery lying upon the ground with its copper (positive) end to the east, and the wire connecting the ends of the battery bent into an arch similar to the course of the sun; the electric current will thus flow up from the east end of the battery, and descend into it at the west end. The fluid that is decomposed by a current of electricity passing through it is termed by Faraday an *electrolyte;* the elements liberated by this decomposition he terms *ions*, distinguishing those liberated at the cathode as *cations*, which in sulphate of copper would be the metal, and those liberated at the anode as *anions*, which would be the acid portion of the sulphate of copper.

The late Professor Daniell, disapproving of the terms cathode and anode, substitued *platinode* for the negative, and *zincode* for the positive, pole. We think these terms are better adapted for electrometallurgy than cathode and anode, which have no direct reference to ordinary conditions; while zincode distinctily expresses the substance dissolved, and platinode the element not acted upon.

Professor Graham adopts the terms *zincous* and *chlorous* poles, as synonymous with zincode and platinode, or positive and negative.

Although the terms positive, negative, and pole, may not be the best, still, under all the conditions of electro-metallurgy, we deem them as appropriate as any of the proposed substitutes, some of which are based on supposed conditions which have not been proved, and may be found incorrect.

When we shall have occasion to use the two terms *pole* and *electrode*, these will be used synonymously: positive and negative electrode are synonymous with positive and negative pole.

*Electrolyte* will be applied to a solution when undergoing decomposition by the electric current passing through it.

The *positive electrode*, or pole, is that metal in the electrolyte which is being dissolved, or, if not capable of being dissolved, at which the acid or solvent of the electrolyte is being liberated, as when sulphate of copper forms the electrolyte, the sulphuric acid is liberated. The *negative electrode*, or pole, is that metal or substance in the electrolyte upon which the metal is being deposited by the influence of the electric current, such as a medal upon which copper is being deposited in an electrotype process.

## BATTERIES.

Single Pair of Plates.—If a piece of ordinary metallic zinc be put into dilute sulphuric acid, it is speedily acted upon by the acid, and hydrogen gas is at the same time evolved from its surface, having a disagreeable smell arising from impurities contained in the zinc or acid. If the zinc be taken out, and a little mercury be rubbed over its surface, an amalgamation takes place between the two metals: the plate becomes of a beautiful bright silver appearance. If the zinc thus amalgamated be again put into the dilute acid, there is no action, for the mercury retains the zinc with sufficient force to protect it from the acid. If a piece of copper be immersed along with the zinc, and the two metals be made to touch each other, a particular influence is induced among the three elements, zinc, copper, and acid; the acid again acts upon the zinc as if no mercury was upon it, but the hydrogen is now seen to escape from the surface of the copper; this action will go on as long as the two metals are kept in contact. Or if, instead of causing the two metals to touch, a wire be attached to each, and their opposite ends are placed in a little dilute acid in another vessel, the same action will take place between the zinc and copper as when they were in contact; but in this instance, the ends of the two wires which dip into the vessel containing acid will undergo a change: the one attached to the zinc will give off a quantity of hydrogen gas, while the one attached to the copper, supposing it to be also copper, will rapidly dissolve.

Figure 542. Represents the zinc and copper, placed in dilute sulphuric acid, brought into contact; in this experiment, gas will be seen escaping from the copper.

Figure 543. Zinc and copper, placed in dilute acid, and wires attached, which, when connected, will exhibit the same effects as in the first case.

Figure 544. Shows the wires connected by means of a liquid, such as acid and water, sulphate of copper, etc.

Figs. 542 543 544.

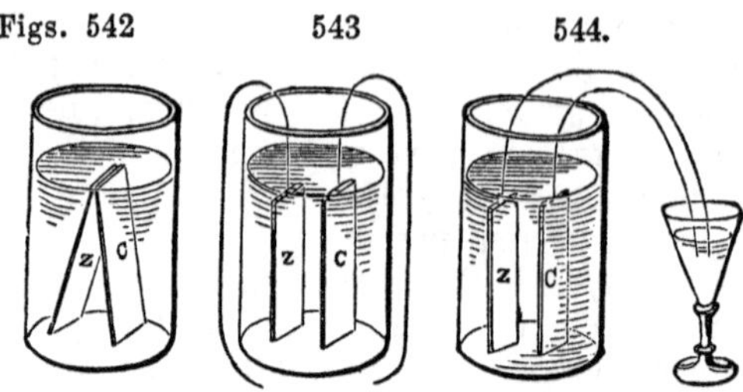


The copper and zinc, c and z, with the acid in the first vessel, Figure 544, constitute a battery of one pair. The wine glass *e*, wi acid, in which the wires are placed, is termed the decomposition cell.

BEST KIND OF ZINC.—The zinc used for the battery should be *milled* or rolled zinc, not thinner than ⅛th of an inch, otherwise the waste will be very great; for amalgamated zinc, when it becomes thin, is so tender and brittle, that the utmost care cannot preserve it whole. The best thickness for the zincs, when their size is upwards of four inches square, is ¼th of an inch; but if under this size, ⅛th to $\frac{3}{16}$th of an inch is the proper thickness. Cast plates of zinc should not be used, as they are negative to rolled zinc, and give less electrical power: they are so porous that no amalgamation will protect them from the action of the acid—producing "local action," as it is termed, which is not only a waste of zinc and acid, but prevents, to a great extent, the production of the quantity of electrical force which the surface of the zinc in use is calculated to give.

AMALGAMATION OF THE ZINC PLATES.—The amalgamation of zinc is a process exceedingly simple; nevertheless, if care be not taken, a very great loss in mercury and zinc is soon effected. A stoneware pan is best to use, and should be sufficiently capacious to allow the zinc plate to lie flat within it: a mixture of eight parts water, and one part sulphuric acid, should be put into the pan, sufficient in quantity to cover the zinc plate, which should lie in it till the surface is perfectly bright. The pan is now raised on the one side, and a little mercury put into the lower part, care being taken that the zinc does not touch the mercury, to prevent which is the object of raising the pan on one side. A little coarse tow, tied to the end of a piece of wood, is dipped into the mercury, which lifts small portions of the metal mechanically, which is then rubbed with considerable pressure upon both sides of the zinc plate, over which the mercury flows easily: the plate is then washed, by dipping it into clean water, and is next made to stand upon its edge in another pan, with two small pieces of wood under it, so as to allow the mercury to drain from it. Instead of tow, an old scratch brush is generally used in plating factories: this is a brush made of fine brass wire, tied upon a piece of wood; but we prefer tow, when carefully employed, as the brass wire amalgamates with the mercury, and causes a loss of that metal. After the zincs have drained for a few hours, the process should be repeated, only it is not necessary to allow the metal to lie in the acid in the second process previous to rubbing in the mercury: after draining a few hours the second time, amalgamation is completed. In the first process a plate, of a foot square, amalgamated on both sides, will retain three ounces of mercury; but for the second process, or any time after, the same size of plate will only retain 1½ ounces of mercury.

Zinc rapidly absorbs mercury, which permeates the whole metal. If the mercury was in quantity, the zinc would dissolve in it; hence the propriety of rubbing the mercury into the zinc, only in small portion, for if allowed to imbibe as much as it is capable of doing, it would not only be a loss of mercury, but the plate would become exceedingly brittle. When too much mercury is used, a portion of it will filter out from the plate by standing, but it carries with it some zinc dissolved, which tends to deteriorate the quality of the metal for the battery.

The zinc in the battery, after being used, should never be allowed to lie in the acid when the battery is not in use, but should be taken out, and the surface carefully brushed with a *hard* hair brush, in water, and then laid by in a safe place. The matter thus brushed off, being an amalgam of zinc, is to be carefully collected, and kept in dilute sulphuric acid, or in the waste acid from the batteries. most of the zinc in this amalgam will dissolve out, so that a great portion of the mercury may be recovered, or by placing these brushings in a coarse cloth bag, and subjecting it to pressure in a screw-press, most of the mercury may by this means be recovered.

ECONOMY IN AMALGAMATION.—If the battery is to be used seldom, and only for a short period at a time, another method of amalgamation may be adopted. The zinc plate, after lying in the dilute acid till the surface is bright, may be rubbed over with a solution of nitrate of mercury, which gives a very thin amalgamation; but this method is unsuitable if the battery is to be in use for several hours together.

When a battery is being worked daily, it will be advisable to repeat the amalgamation from time to time, otherwise local action will begin, and the working power of the battery be weakened, while the loss in zinc will be increased.

The following is the proportional rate which we have found on the large scale under the most favorable circumstances. A new zinc plate, amalgamated as described, working continuously—

24 hours, zinc lost 12½ ounces—copper deposited 12 ounces;
48 hours, zinc lost 20½ ounces—copper deposited 17 ounces;
60 hours, zinc lost 34 ounces—copper deposited 24½ ounces.

From these and similar data we found that the most economical way of using zincs is the following: after being in the battery twenty-four hours, they are to be taken out, brushed, and laid aside; after working other twenty-four hours, they are to be again brushed and *immediately* re-amalgamated: if these directions are attended to, ½ ounce of mercury will be sufficient for one foot square of zinc, both sides.

The advantages of proper amalgamation will be made more evi dent in the sequel. We have only to add here, in consequence of an oft-expressed fear of the danger of working with quicksilver, that no apprehension need be felt: the skin does not absorb it, and there being no heat required in the operation that could convert the mercury into vapor, the only state in which it is dangerous, no salivation can take place.

Distance between the Battery Plates.—To return again to the battery-cell. It will be found that if the two metals—the zinc and copper in acid, Fig. 544—be put very close to each other, the action will be much more rapid than when they are far apart. It will also be found that, allowing the zinc and copper to be kept at one distance, but the wires in the decomposition-cell to be put at different distances, similar results will take place. When the wires are close the action in the battery-cell will be more powerful than when the two wires are put farther apart: these properties are ap plicable to all batteries and decomposition-cells of every kind. The following results will give an idea of the relations of these several conditions:

1st. One pair of copper and zinc plates, measuring superficially 6 square inches, were immersed in a solution consisting of 1 acid to 35 water: plates of copper of equal size to those of zinc and copper were laid in the decomposition-cell, which was then filled with a liquid of equal strength to that in the battery-cell; the plates in the battery-cell and the decomposition-cell were then placed one inch apart: in four hours

The zinc in the battery-cell lost by dissolving 10½ grains;
The copper dissolved in decomposition-cell 10 grains.

2d. The battery-plates were put 12 inches apart, and the plates in decomposition-cell 1 inch apart: in four hours

There were dissolved in the battery-cell, zinc 7 grains;
In decomposition-cell, copper 6 grains.

3d. The battery plates were placed 1 inch apart, and the plates in decomposition-cell 12 inches apart: in four hours

The zinc in battery-cell lost 4½ grains;
The copper in decomposition-cell lost 3½ grains.

These results show the importance of attending to the condi tions of the respective agents, and also, that distance in the decom position-cell offers greater resistance than distance in the battery cell.

Different Elements of Batteries.—Although our observa tions have been made on zinc, copper, and dilute sulphuric acid in the battery-cell, still these are not the only essential elements in a battery, as almost any two metals with a liquid similarly arranged will produce an electric current; but the current will vary according to the nature of the metals employed, and the effects produced upon them by the solution in which they are placed. If the exciting solution has the power of acting upon both metals, as when zinc and copper are immersed in dilute nitric acid, the current of electricity produced by the action of the acid upon the zinc will be neutralized to an extent corresponding to the relative action of the acid upon the copper. To have any effective electrical power, it is necessary that one of the metals employed be capable of combining easily with one of the elements of the solution in which they are placed, and forming a soluble salt, while the other does not; and the power obtained under proper circumstances has an intimate relation with these two properties in contrast. The metal which undergoes solution is termed the positive metal, the other the negative metal. Metals are not considered to possess any intrinsic negative or positive principle; their relations in this respect are governed solely by the circumstances in which they may be placed. For instance, if we connect a piece of copper and a piece of iron, and immerse them in acidulated water, the iron is dissolved, and is positive in relation to the copper; but if the same metals are immersed in a solution of yellow hydro-sulphuret of potassium, the copper is dissolved, and is positive relatively to the iron. Hence, to obtain a galvanic battery, the conditions are simply to provide two metals, and immerse them in a solution capable of acting upon the one and not upon the other. The first table shows the order in which the common metals stand to each other, in respect of their relative negative and positive properties, when immersed in water acidulated with sulphuric acid. The second table is given by Gmelin as the relations of the metals, in water and sea water, the most intensely negative metal standing highest, and the metal which acts most positively standing lowest:

| | |
|---|---|
| Platinum | Platinum |
| Gold | Gold |
| Antimony | Silver |
| Silver | Copper |
| Nickle | Bismuth |
| Bismuth | Antimony |
| Copper | Iron |
| Lead | Tin |
| Iron | Lead |
| Tin | Cadmium |
| Cadmium | Zinc |
| Zinc | |

According to this arrangement, each metal is positive with respect to all that stand before it, and the electrical conditions of any pair become the more contrasted the further apart they stand in the scale. Thus, a battery composed of zinc and platinum is much more powerful than one composed of zinc and copper; and again, copper and iron make a very weak battery.

A battery may also be formed by having one metal and two kinds of solutions, separated by a porous diaphragm. For example, we may have strong nitric acid in one division, and dilute sulphuric or muriatic acid in the other; and by putting into each a piece of clean iron, a powerful current is obtained. These and several other arrangements of solutions and metals, are expensive and troublesome to keep in order, and are therefore never used for practical purposes in the art of electro-metallurgy.

PROPERTIES OF METALS FIT FOR BATTERIES.—In looking to the above table, it may be asked, "Since lead stands next to copper, and is so much cheaper, why should it not be used instead?" The reason is, that there are other properties which a metal, especially that used as the negative element, ought to possess to fit it for use in a voltaic arrangement; such as the power of freely conducting an electric current, of keeping a bright surface, and not becoming oxidized; none of which properties belong to lead. Could that metal be kept from oxidizing, a very powerful current of electricity might be obtained by using it with zinc; but its surface soon gets coated with an oxide possessing none of the properties of the metal, and hence the arrangement becomes zinc and oxide of lead, which produces but a weak current of electricity. These remarks refer to any metal that is subject to oxidization—an incident which is often a source of annoyance to the electrotypist when using copper plates.

Lead slightly amalgamated, and used as the negative metal with zinc, produces a very constant current for a time.

Lead is also a very bad conductor of the electric current, which renders it unsuitable for an element in the battery, the negative metal being considered as only acting the part of a conductor: this property materially affects the available power of an arrangement.

The following table shows the *relative Conducting power* of the respective metals:

| | |
|---|---|
| Silver | 120 |
| Copper | 120 |
| Gold | 80 |
| Zinc | 40 |
| Platinum | 24 |
| Iron | 24 |
| Tin | 20 |
| Lead | 12 |

We have just stated that a battery composed of zinc and platinum will be more powerful than one composed of zinc and copper, so far as regards their negative and positive tendencies; but so much does the conducting power of the negative metal affect the practical usefulness of a battery, that notwithstanding the fact that platinum is much more negative than copper, there is so much of the effective electricity expended in overcoming the resistance which the inferior conductibility of the platinum offers to the progress of the current, that a battery of zinc and copper proves to be a more effective and useful battery for electro-metallurgy than one made of zinc and platinum. Hence also the reason that iron and copper, or iron and any other metal, make but an indifferent battery —iron being a bad conductor; while lead, which will be seen in the table, stands lowest in this property, is therefore unfit for batteries.

In fitting up a voltaic arrangement with a negative metal that is not a good conductor, such as platinum, the closer it is placed to the exciting liquid, in connection with another metal that is a good conductor, the better; because the current obtained will be the more effectual.

The following experiments will illustrate these remarks with a few of the common metals used as negative electrodes. There were, in each battery, six square inches of each metal exposed to the action of the acid, which was sulphuric acid diluted with 25 parts of water. The poles were of the same size, of copper, placed in sulphate of copper; and the quantity of copper deposited was taken as the data, each trial being of a different length of time:

FIRST, IN ACTION HALF AN HOUR.

| BATTERY. | DEPOSITED. |
|---|---|
| Tin and zinc . . . . | 1·7 grains. |
| Copper and zinc . . . | 1·8 " |
| Platinum and zinc . . . | ·5 " |
| Platinized silver and zinc . | 2·0 " |

SECOND, IN ACTION TWO AND A-HALF HOURS.

| | |
|---|---|
| Tin and zinc . . . . | 5·9 grains. |
| Copper and zinc . . . | 8·8 " |
| Platinum and zinc . . . | 4·5 " |
| Platinized silver and zinc . | 9·7 " |

THIRD, IN ACTION SIXTEEN HOURS.

| | |
|---|---|
| Tin and zinc . . . . | 64·5 grains. |
| Copper and zinc . . . | 50·5 " |
| Platinum and zinc . . . | 43·0 " |
| Platinized silver and zinc . | 67·3 " |

From these few experiments it appears that tin and zinc, when used for a long time, constitute a very effective battery. It is very constant in its action, and thus suited for time. It stands next to platinized silver. The whole, in nineteen hours, gave respectively—

| | | |
|---|---|---|
| Platinized silver . . | 79· | grains. |
| Tin . . . . | 72·1 | " |
| Copper . . . . | 61·1 | " |
| Platinum . . . | 48·0 | " |

BABINGTON'S BATTERY.—If we look back to the description given of the voltaic pile (page 489), and the improvement made upon it by Cruikshanks, we perceive the relation they bear to the pieces of copper and zinc mentioned in page 510; but the relation is more apparent in Babington's improvement upon Cruikshank's battery. When working with this battery, it was found that the energy of

Figs. 545 546.

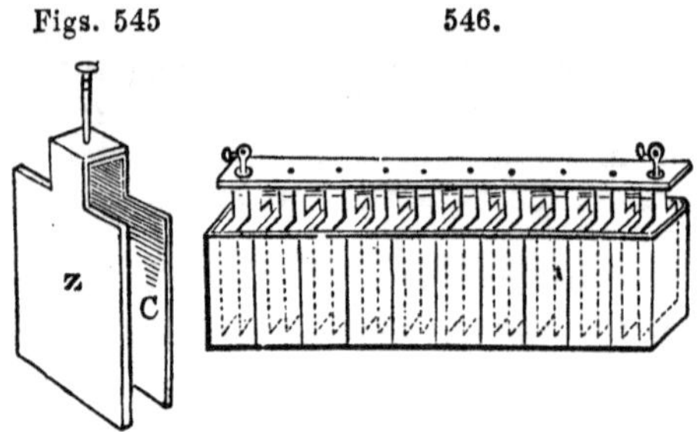


the battery did not depend, as was supposed, upon the extent of surface of the zinc and copper which were in contact, but upon the extent of surface of these metals in contact with the liquid with which the battery was excited; and that it was sufficient if the zinc and copper touched each other in a single point;—provided that the plates were plunged into the liquid, and that the copper plate should be exactly opposite to a zinc plate in the same cell, a space between them. Hence, instead of soldering the zinc and copper together, as Mr. Cruikshanks did, it was enough to effect a communication by turning over a portion of the copper plate at the top, and soldering it to the upper extremity of the zinc. Thus—C, the copper, is bent over to touch and be soldered to the zinc plate Z. For this arrangement, the wooden trough was divided, by plates of glass or varnished wood, into as many cells as there were pairs of zinc and copper. The cells being filled with the acid, or exciting solution, the metals were then placed into them in such a manner that each pair of

Fig. 547.

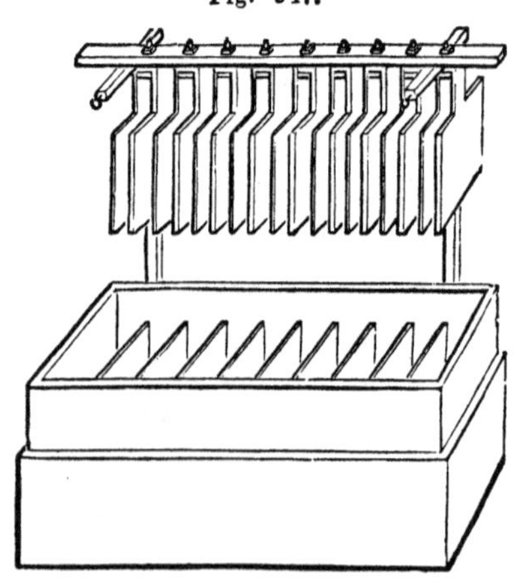

zinc and copper plates had a partition between. By this arrangement the zinc of one pair faced the copper of the next pair in the cell, as shown in Figures 546 and 547. The former represents the plates immersed in the solution; the latter, the plates suspended on a rack over the solution. This arrangement was termed Babington's Battery.

WOLLASTON'S BATTERY.—Although we have spoken of the great value of amalgamated zinc for batteries, still at the period when the arrangement just described was introduced, amalgamation was not known; and the zinc plates were, therefore, always liable to be destroyed by the acid. It was, consequently, of importance that no zinc should be exposed to the action of the acid that was not calculated to give electricity, as the energy of each pair of plates depends upon the extent of surface of the two metals *exactly opposite to each other.* It will be evident that in Babington's arrangement only one side of the zinc was effective in giving electricity, while both sides were exposed to the action of the acid. To obviate this defect, Dr. Wollaston caused the copper plate to *surround the zinc,* by which the whole surface of the zinc exposed to the acid was made effective in producing electricity, and thereby doubling its quantity without further cost. The accompanying figure (Fig. 548) shows the manner in which this battery was originally constructed.

Fig. 548.

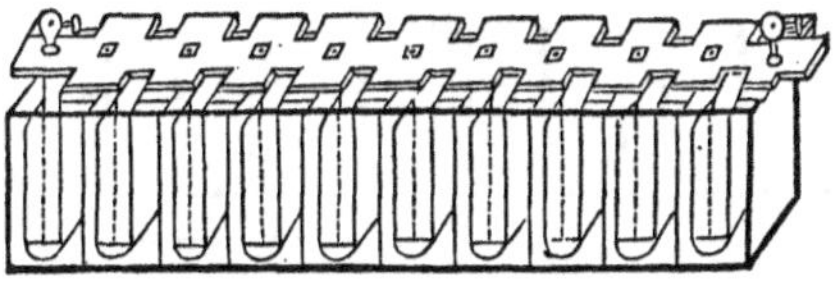

This improvement, if we except several modifications of construction for the facility of taking the plates asunder for cleaning, etc., did all for this kind of battery that could be done.

MODIFICATION OF WOLLASTON'S BATTERY NOW IN USE.—Wollaston's battery is still generally used in large factories for depositing metals; and it is found by experience to be the most convenient and economical of all the batteries yet contrived. The modification we have found to be very suitable, and practically useful, may be thus described. In the arrangement represented above, when amalgamated zincs are used, small quantities of amalgam fall from the zinc plates upon the copper, which not only occasion local action, but the mercury amalgamates with the copper, spreads over it, and to a great extent lessens its efficiency; and as the copper must be red hot to expel the mercury, much loss of copper as well as mercury is the result. To obviate this defect, the copper is connected above the zinc and left open at bottom; as, for example, a thin sheet of copper, of dimensions according with the size of the cells in the battery, is cut thus: This copper is bent in the middle at *b*, the ends *a a* dip into the cells, while

Fig. 549.

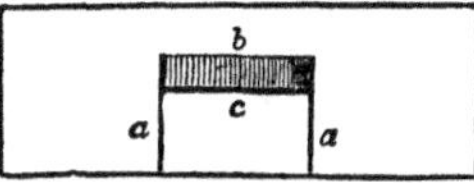

c, is bent over to connect with the zinc plate of the neighboring cell, thus:

Figs. 550 551.

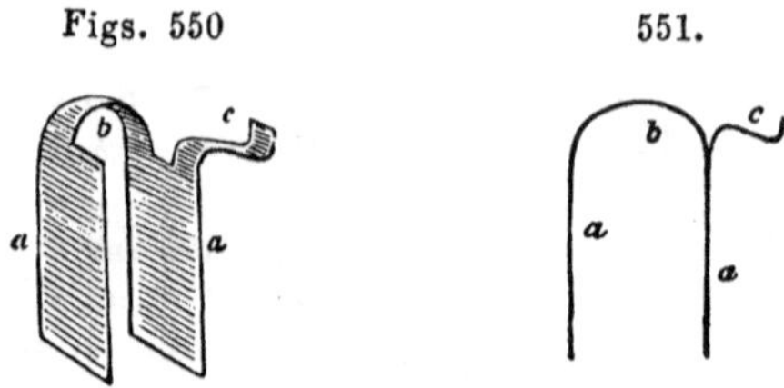


The zinc plates are placed between the bent copper *a a*. The following diagram of a battery of several pairs of plates will illustrate these observations:

Fig. 552.

Z Z Z Z, The zinc plates.
C C C C, Copper plates.
P P P, Partitions of trough, which are generally made of thin wood.
*w w*, Wires from battery.

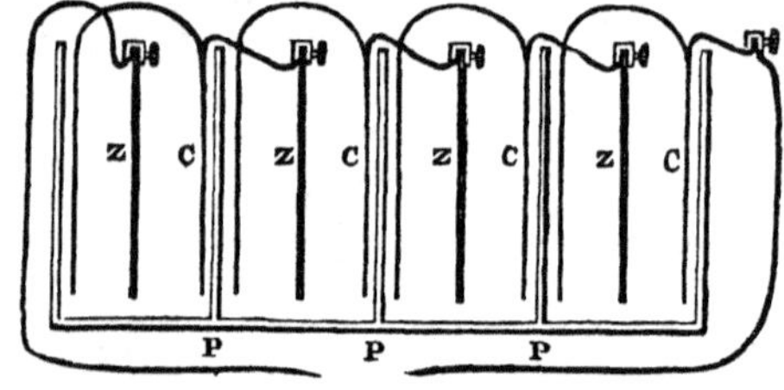


The zinc and copper are connected together by a binding screw. To construct this battery the zinc plates are put in first, being made to slide in grooves, cut in the sides of the trough, the plates standing in the centre of their respective cells: the copper plates are put in, and the copper bands marked *c* are made fast to the zincs by binding screws, care being taken that the parts where they are connected are clean and bright, and that the copper and zinc touch nowhere else. A battery of nine pairs of plates can be fitted up and made ready for action in ten minutes.

In fitting up batteries of this sort, we are aware that sometimes great care is taken that the partitions in the trough be perfectly water-tight, and also formed of some non-conducting material, such as glass, or of wood, either pitched or saturated with some non-conducting substance; but we have found in practice that these precautions are not required, the principal thing to attend to being, that the metals should not be allowed to touch, except in their proper connections.

Defects of Common Acid Batteries.—Although we have spoken thus favorably of the principles upon which Wollaston's battery is constructed, still as a philosophical instrument it is far from being perfect: hence the many modifications of it which have been recommended. Indeed, electro-chemists, since the time of Volta, have been endeavoring to invent an instrument free from the defects which attach to Wollaston's—one capable of giving, at the same time, a constant and powerful current, abundant in quantity, and of great intensity. The success and results of these endeavors are so closely connected with the art of electro-metallurgy, and the knowledge of them is so essential to a successful prosecution of the art, that we must not be sparing in our descriptive details.

In operating with a Wollaston's battery, or any other arrangement composed of similar elements, such as zinc, sulphuric or muriatic acid, and copper, silver or platinum, it will be found that the current of electricity obtained diminishes in quantity and strength in proportion to the time of action. This is the result of various causes:

1st. The hydrogen which is evolved at the surface of the negative metal in the battery, which we shall say is copper, adheres with considerable force to the surface of the metal, and consequently obstructs its superficial influence, so that the quantity of electricity which the surface of the two metals is calculated to give is much lessened.

2d. After the battery is in action a short time, a portion of the sulphate or chloride of zinc, formed in the battery by the solution of the zinc, becomes reduced upon the surface of the copper. This reduction is supposed to be owing to the electrolyzation of the zinc solution by the passage of electricity, but it is, more probably, caused by a galvanic action upon the copper plate and the solution in the battery. It generally begins at the lower edge of the copper plate, and spreads upwards. This weakens the electric current, both by inducing a galvanic action between the zinc and the copper, upon which it is deposited, and by its tendency to send a current of electricity in an opposite direction to the main current, thereby neutralizing to a great extent the original power of the circle.

3d. When copper is used it becomes gradually covered over with a thin, black, slimy coating of oxide and other impurities, which materially affects the regularity and strength of the current: this is a source of considerable annoyance in working, and necessitates a regular cleaning of the coppers, which should be done immediately on being taken out of the battery, by brushing with a hard hair brush in water; but when the battery has been long in action, this mode of cleaning is insufficient: the plates will then require to be rubbed over with a little dilute nitric acid, and then washed. If the black coating be allowed to dry upon the coppers, they must then be dipped into strong nitric acid till their surfaces are acted upon; or they may be moistened with a little urine, then brought to a dull red in the fire, and immediately plunged into water; but in both cases there is a loss of copper. A small quantity of the black matter, upon being tested, gave oxide of copper, with a trace of iron, antimony, and lead, which are the general impurities of sheet copper.

Max, Duke of Leuchtenberg,* has giving the following results

---

* Progress of General Science, vol. ii.

of an analysis of the black matter found upon the copper plate forming the positive electrode in a copper solution.

| | |
|---|---|
| Sand . . . . . . . | 1·90 |
| Antimony . . . . . . | 9·22 |
| Tin . . . . . . . | 33·50 |
| Arsenic . . . . . . | 7·40 |
| Platinum . . . . . . | 0·44 |
| Gold . . . . . . . | 0·98 |
| Silver . . . . . . . | 4·45 |
| Lead . . . . . . . | 0·15 |
| Copper . . . . . . | 9·24 |
| Iron . . . . . . . | 0·30 |
| Nickle . . . . . . | 2·26 |
| Cobalt . . . . . . | 0·86 |
| Vanadium . . . . . . | 0·64 |
| Sulphur . . . . . . | 2·46 |
| Selenium . . . . . . | 1·27 |
| Oxygen . . . . . . | 24·82 |
| | 99·98 |

An analysis of this sort invests the subject with great interest. We wish there had also been given an analysis of the copper that was used as the electrode, and the quantity of black matter obtained, and the quantity of copper dissolved to yield that product: for the analysis indicates an amount and diversity of impurities in copper that has never hitherto been thought of. Both the number and the proportions are startling. However, it is a known fact, that the more impure the copper is that is used for a pole, the black matter is not only more easily, but more abundantly produced.

Another source of weakness to the electric current, and which affects more or less all batteries of whatever construction, arises from the action of the acid upon the zinc. The more freely this action is allowed to proceed the more constant and powerful is the battery. The acid in combining with the zinc forms a salt, which, if it adhered to the surface of the plate, would soon stop further action; but this salt being soluble in water, is dissolved from the surface of the plate as soon as formed, allowing a new surface to be exposed. But water can only dissolve a certain quantity of the salt, and its power of dissolving decreases as it approaches to the limit of saturation: hence there is a constant tendency to a decrease of power in the battery, and if means be not taken to withdraw the salt of zinc formed, the battery will continue to decrease in power, till at length it ceases to act. But long before the battery ceases to act, the presence of sulphate of zinc manifests itself in several ways, neutralizing the efficacy of the battery. The zinc salt, as it dissolves from the plate, being heavier than the acid solution, falls to the bottom; hence in a very short time the solution as formed of strata of different densities, and this induces a galvanic action between the lower and upper portions of the plates, both copper and zinc, and accounts for the deposition of zinc on the bottom part of the plates, as above referred to. This local galvanic action between the bottom and top parts of the zinc plate is sometimes so great when the battery has been long in action, as to double the thickness of the zinc plate at bottom, while the part near the surface of the solution is nearly penetrated by the acid; and when a battery is formed of a number of pairs, the terminal zincs are those most affected, the one forming the negative terminal or pole more so than the other. We have found a deposition of 6½ ounces of zinc upon the two lower inches of a plate terminal, which measured, in the solution, six inches by five, the battery having been in operation but eighteen hours. When this occurs, the quantity of electricity circulating through the battery is very small. Although this evil may not proceed to the extent of having quantities of zinc deposited upon the bottom part of the plates, still the tendency to deposition which every one who employs a battery must have observed, as also the more rapid action of the acid on the upper parts of the plates, shows that the action of the acid over the surface of the plate is very irregular, and consequently the quantity of electricity must be irregular in the same degree, often producing in the battery an intermittent action.

Various means have been devised for removing the sulphate of zinc, and adding corresponding quantities of new acid water; the most simple and effective of which, according to our experience, is to make the battery trough much deeper than is required for the plates, which may be supported either by grooves in the side of the trough, cut to the proper depth, or by a fillet of wood, or perforated false bottom; so that the zinc salt when formed may fall under the plates, and thus a much longer time elapse before its presence produces any decidedly bad effect.

There can be no doubt, we think, that some easy means will yet be devised for carrying off the dense solution of sulphate of zinc, before it rises to the plates, and for replacing it by acid water from above, thus giving to the battery a uniformity and steadiness of action it does not at present possess. There have been many ingenious contrivances tried for this purpose by the amateur, but we have seen none so simple and economical for manufacturing purposes, as that referred to.

DANIELL'S BATTERY.—A few years ago, some of the disadvantages now detailed were to a great extent overcome by a very ingenious arrangement discovered by the late Professor Daniell. The discovery consists in the separation of the zinc from the copper by a porous diaphragm, such as bladder, unglazed porcelain, etc., and the use of two distinct fluids. The portion of the battery containing the zinc is charged with dilute acid as before, but the portion containing the copper is filled with a solution of sulphate of copper. The action in this battery is similar to that described in the ordinary battery; the zinc is dissolved by the acid, but the hydrogen, instead of being evolved at the copper plate, combines with the acid of the sulphate of copper: metallic copper is thus set at liberty upon, and combines with, the copper plate of the battery, not only maintaining but improving its surface, during the evolution of a constant current of electricity. From the constancy of the current maintained the battery has been termed the *Constant Battery*. The construction of a single pair is described by Professor Daniell in the following terms:

Fig. 553.

"A cell of this battery consists of a cylinder of copper 3½ inches in diameter, which experience has proved to afford the most advantages between the generating and conducting surfaces, but which may vary in height according to the power which it is wished to obtain. A membranous tube, formed of the gullet of an ox, is hung in the centre by a collar, and a circular copper plate, resting upon a rim, is placed near the top of the cylinder, and in this is suspended, by a wooden cross-bar, a cylindrical rod of amalgamated zinc, half an inch in diameter; the cell is charged with eight parts of water, and one of oil of vitriol, which has been saturated with sulphate of copper, and portions of the solid salt are placed upon the upper copper plate, which is perforated like a collander for the purpose of keeping the solution always in a state of saturation. The internal tube is filled with the same acid mixture without the copper. A tube of porous earthenware may be substituted for the membrane with great convenience, but probably with some little loss of power.*

A number of such cells may be connected very readily, by attaching the zinc of the one to the copper of the other, and (as shown in Fig. 554) thus forming an intensity arrangement of great power and constancy.

Fig. 554.

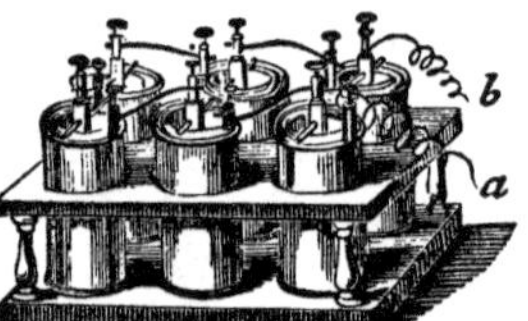


This arrangement of battery is eminently suited to all kinds of electrical operations, and it may be borne in mind that it was by operating with this battery the idea of electro-metallurgy first occurred. In this battery we see that the evils arising from the slow liberation of the hydrogen from the surface of the negative metal, and the deposition of the zinc upon the copper, and also the blackening of the surface of the copper, are all surmounted. Nevertheless it is not used to any extent in the art of electro-metallurgy, it being much less economical than the ordinary batteries, from the quantity of copper salt necessary to keep it in a work-

* Daniell's Chemical Philosophy, 2d edition, 1843, p. 504.

ing condition, and from the necessity of using porous diaphragms, which speedily wear out. If the diaphragm is made of animal membrane, the acid very soon destroys it; and although unglazed porcelain lasts a little longer, the acid acts upon the alumina, so that after a few days' working the diaphragm becomes too porous; and if the zinc plate touches the porous vessel, a circumstance very difficult to avoid, there is very soon formed in and upon the porous surface a deposit of copper which speedily renders the cell useless, besides producing a loss of copper. The saturation of the zinc solution, already spoken of, not unfrequently produces the same effects—the saturated portion of the bottom becomes reduced by the local action, and thus often a minute point of metallic zinc touches the cell, and forms a nucleus for a deposit of copper upon the porous cell, which spreads over the surface very rapidly. There are always pieces of amalgamated zinc, like fine scales, falling to the bottom of the cell, which also form nuclei for the deposition of copper upon the porous cell.

After the cells have been some time in use, if they are laid aside and allowed to dry, they are very liable to break. Care should be taken to keep them in clean water till the salts within the pores are dissolved out; but if this precaution is taken they may be preserved for a long time if only used occasionally.

The remarks upon the economy of the arrangement just described have reference to its use as an instrument, or separate battery, for the deposition of a metal in a separate cell (decomposition cell); but not to the arrangement known in electro-metallurgy as the *single cell* process, which is simply a modification of one pair of Daniell's arrangement; a description of which, with its comparative economy, will be given in another part of this treatise.

Professor Daniell says that any depth of cell may be used according to the power required, but this cannot be done with equal advantage, for a large surface of zinc in a cell is not the most economical, when a great quantity of electricity is required.

GROVE'S BATTERY.—Another battery, constructed upon the same principle as Daniell's, but differing in the arrangement of the metals, and the substances used to excite them, was invented by Mr. Grove, and is known as *Grove's Battery*. In this arrangement, platinum is used instead of copper, and strong nitric acid instead of the sulphate of copper of Daniell's battery. One pair may be fitted up conveniently in a tumbler or jelly-pot. A cylinder of zinc is placed inside the tumbler; within this cylinder is placed a porous vessel, in which is a slip of platinum either in sheet or foil; the porous vessel is filled with strong nitric acid, and the tumbler with dilute sulphuric acid: a wire is next attached to each metal, and the battery is complete.

Fig. 555.

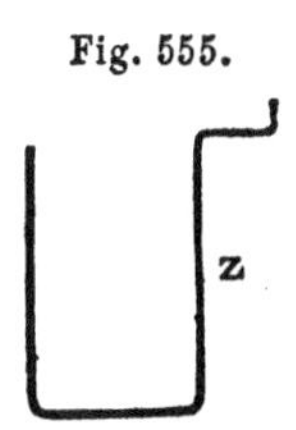


When a series of pairs is to be used, the form we have found most convenient is to arrange the metals in the same manner as we have described for Wollaston's trough (page 518). The zinc is formed in the same shape as shown by Fig. 555. The zinc is placed in the cell of the trough, and the porous vessel which should be flat is placed within the zinc, so that the platinum in it may be connected with the zinc of the neighboring pair, as represented in figure 556.

Fig. 556.

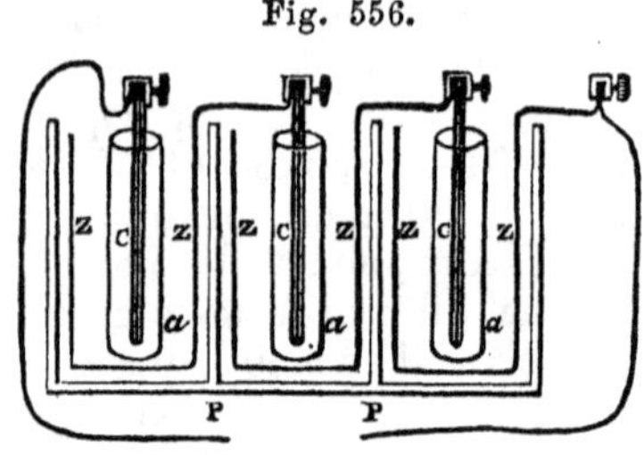


zzz, Are the zinc plates of the form of figure 555.

*aaa*, Porous cells filled with nitric acid.

ccc, Plates of platinum united to the zinc at top by binding screws.

PP, Are partitions. The divisions of the battery trough need not be water tight, but merely such as will prevent the zincs from touching one another.

It will be seen that by this means any number of pairs may be easily arranged. Care, however, must be taken, when fitting up such an arrangement, that the platinum be kept closely connected with the zinc by a large surface, otherwise the platinum will be fused at the connections. A flat piece of wood, with a groove to fit the zinc, is often made the means of keeping the two metals together, but we prefer flat binding screws of brass, for if kept clean they assist the connection, being good conductors. The fusion of the platinum connections, a practical and often expensive annoyance, may, however, be completely prevented by coating about half an inch of the end of the platinum, either with copper or silver, which is easily effected by the electro-process: the coated part is then connected with the zinc by any convenient means without the risk of fusing.

Figure 557 represents a section of Mr. Grove's nitric acid battery, in a series of four pairs of zinc and platinum. The outer thick line, A B C D, is an earthenware trough, which is divided into four cells. The dotted lines represent four porous vessels, of a size sufficient to contain about double the quantity of liquid that is contained between the outer surfaces of the porous vessels and the earthenware cells. The dark central lines are the plates of amalgamated zinc; and the thinner lines, that bend round under the porous cells, show the position of the platinum foil, which is attached to the zinc plate by small screws.

This form of battery is also free from some of the objections to the common battery, but it is seldom or never used in the ordinary processes of electro-metallurgy. Its advantages over every other known form of battery are, its great activity of action, and intensity or power of current—a circumstance not generally sought after by electro-metallurgists. But if it were duly considered that a battery consisting of three pairs of zinc and platinum is far more effective than an ordinary battery of ten or twelve pairs (although the elements of its construction are more expensive), it would stand a fair chance of being adopted as the more economical battery of the two.

Fig. 557.

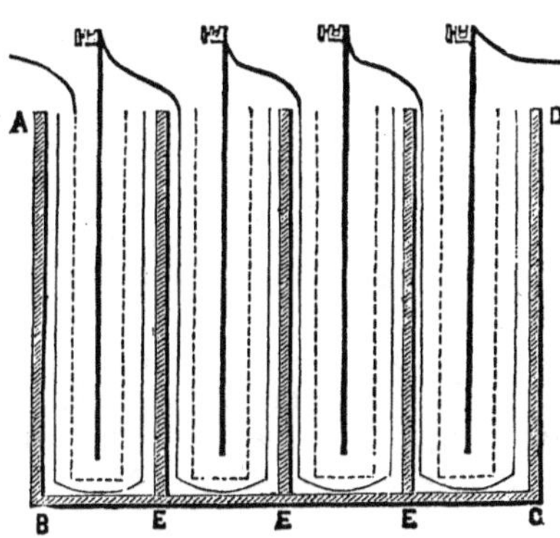


The porous cells have not the objection of being closed up by the deposition of metals upon or within them; but they are affected by the acids, and by long working they become too porous, the nitric acid passing through and causing rapid destruction of the zinc. It wants the constancy of Daniell's, its quantity declining rapidly when long in action—one feature we have often experienced which we have not seen observed by other experimenters. When working with a Grove's battery of from eight to twelve pairs, the platinum being six inches by seven inches, after the battery was in action for four or five hours, during which it diminished in quantity gradually, all of a sudden it seemed to recover its energy, giving a quantity and power not much less than it gave in the first hour, and then declined again rapidly, but occasionally renewing its vigor for short periods. We have thought it probable that this reaction may be caused by the formation of nitrate of ammonia in the rapid decomposition of the nitric acid during the first action of the battery, which, as it accumulates, may be reacted upon.

The prevailing and permanent objection to the use of this arrangement, not only for manufacturing purposes, but for many experimental purposes, is this, that it emits nitrous fumes which corrode every thing within their reach, and prove very disagreeable to every person breathing them. For a small single pair of Grove's battery it is very convenient to use a circular form, in which case a little cylinder of wood, bored so that its sides be about ¼th of an inch thick, does well for a porous cell, and will last a long time.

Bunsen's Battery is a modification of Groves' and much used on the continent for electrical purposes. Carbon is used instead of the platinum, and for convenience it is generally made cylindrical. The mode of construction is to form a hollow cylinder by coking pounded coal in an iron mould, then soaking the coke cylinder in a solution of sugar and calcining a second time, which gives great compactness. The porous cell containing the zinc and dilute acid is placed in this cylinder, and the whole put into a glass or stoneware vessel charged with nitric acid; of course the coke cylinder and zinc are connected, and thus the battery is completed.* Nitrous fumes are evolved from this battery also.

SMEE'S BATTERY.—Some of the defects in the common battery of zinc and copper were much lessened by an ingenious contrivance of Mr. Alfred Smee. This gentleman had observed that if the copper plate of the battery be roughened, either by corrosive acids or by rubbing the surface with sand paper, its action was made much more efficient, the rough surface evolving the hydrogen much more freely. Taking advantage, therefore, of this principle, he covered platinum foil with a finely divided black powder of platinum, deposited by electricity from a solution of that metal, and used this in place of the copper in the ordinary battery. Instead of platinum foil, Mr. Smee soon after adopted silver foil, which is much less expensive. The method of preparing these plates is given by Mr. Smee as follows:—"The silver to be prepared for this should be of a thickness sufficient to carry the current of electricity, and should be roughened by brushing it over with a little strong nitric acid, so that a frosted appearance is obtained. It is then washed and placed in a vessel with dilute sulphuric acid, to which a few drops of nitro-muriate of platinum has been added. A porous tube is then placed into this vessel with a few drops of dilute sulphuric acid; into this tube a piece of zinc is put, contact being made between the zinc and silver; the platinum will, in a few seconds, be thrown down upon the silver as a black metallic powder. The operation is now completed, and the platinized silver ready for use."† A simple method, which obviates the use of a battery is thus described: lay the silver between two pieces of sand paper, and press it with a common smoothing iron, then pull the silver out while under the pressure. The platinum solution is made very hot, and the silver dipped in it for some time, which effects the coating.

The nitro-muriate of platinum is easily prepared: take one part of nitric acid, and two parts of hydrochloric acid (muriatic acid); mix together and add a little platinum, either as metal or sponge; keep the whole at or near a boiling heat; the metal is then dissolved; forming the solution required.

* A more simple and less expensive modification of this arrangement, is to put a solid bar of carbon or coke into a porous cell filled with nitric acid, which is placed in a stone or glass jar filled with dilute sulphuric acid, having a cylinder of zinc surrounding the porous cell, leaving about one inch of space between the porous cell and zinc, similar to a Daniell's, forming a modification of Grove's, having carbon instead of the platinum.

† Smee's Elements of Electro-Metallurgy, 2d edition, p. 24, 1843.

Several experiments have been tried with a view to substitute a cheaper metal than silver to deposit the platinum upon, but not with much success. Cheap metals have also been coated with silver by the electro-process, and been used for depositing the platinum upon. The most successful is a composition metal made of tin, lead, and a little antimony, rolled into sheet and plated by silver; this was found very convenient, because it could be easily bent into any required shape, and it keeps its place without the necessity of fixing in frames as required by thin silver; nevertheless, for constant work these plates are found not to present any permanent advantage, and have been abandoned; besides, to give a sufficient coating of silver, becomes as expensive as silver foil.

Mr. Smee, in constructing his battery, has been guided by the expense of the silver, and therefore reverses the order of arrangement introduced by Wollaston, by surrounding the platinized silver with the zinc. Fig. 558 represents a single cell of this form of battery. A, is the jar containing the solution. Z Z, the two amalgamated zinc plates. S, the platinized silver plate. The whole are suspended by a cross bar of wood; and as it is essential to the proper working of the battery that the plates be always parallel to one another, the wooden frame is generally extended round the edge of the thin silver plate, though it is not so represented in the figure. One of the clamps at the top of the wooden bar is connected with the platinized silver plate, and the other with a pair of zinc plates. Instead of a glass or stoneware jar, small square troughs, made of gutta percha, are often used for the Smee's battery, as they suit admirably, and are not liable to break.

Fig. 558.

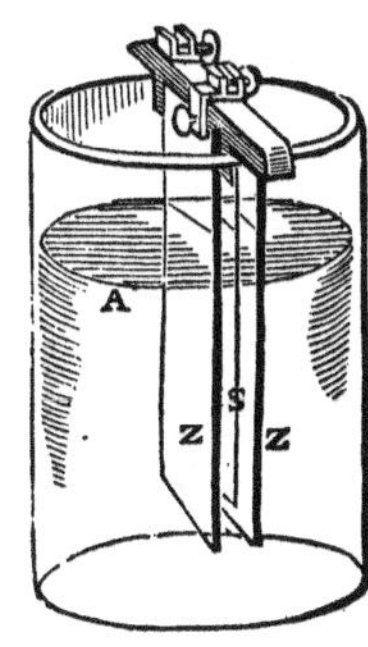


When *intensity* of electricity is required, it is necessary to use a number of such cells, which may be arranged in a wooden frame, in the manner shown by Fig. 559, where *a b* represent the two poles of the battery, and *c c* the wires by which the cells are connected with one another.

Fig. 559

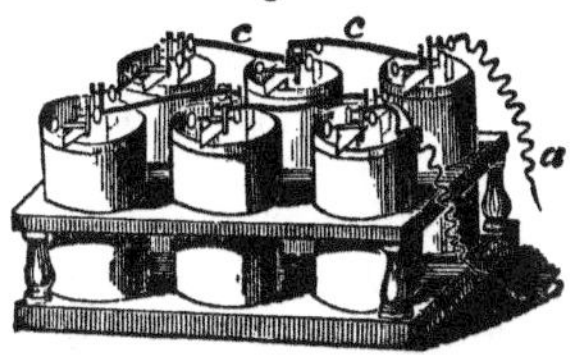


A superior form of compound Smee's battery, contrived by Acland, is represented in Fig. 560. In this apparatus the plates are all connected to a frame that can be elevated or depressed by means of an iron rod and rachet wheel, so that the plates may be either partially or entirely immersed in the solution, or raised at pleasure out of it. The connections are so contrived, that by a slight alteration the battery is adapted to afford either *quantity* or *intensity* of electric power. It is usually made to contain six cells, any number of which can be used at once that a given process may require. The exciting fluid is contained in an incorrodible stoneware trough, placed in a mahogany box. A battery of this description, each silver plate of which measures 20 square inches, has sufficient power, when decomposing water, to disengage one cubic inch of mixed gases in 50 seconds, and will heat to redness 4 inches of platinum wire.

Letter B in Fig. 560 represents an apparatus for showing the de

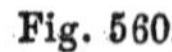

Fig. 560.

composition of water into oxygen and hydrogen gas by the voltaic battery.

It will be observed in Smee's arrangement that there are two surfaces of zinc, in every pair exposed to the acid, which do not give off any electricity, but when long in use are much acted upon, forming a consideration of some value to a manufacturer.

The silver used is very thin, and liable to crack when taken from its frame, and therefore cannot be made into different constructions of battery in the same manner as we can do with copper. It is also liable to have zinc deposited upon its surface when long in action.

There is we believe no arrangement of battery better known and more used by amateur electrotypists than Smee's, and there are probably none better adapted for small operations; but it has not been introduced to any extent in the factory. When used in series, the advantages it possesses over Wollaston's do not counterbalance the extra labor and expense attending its use, and many who have tried it in the operations of the factory have for these reasons given it up.

Numerous modifications of these different batteries have been proposed from time to time, intended for different objects, but those given embrace all that are used for electro-metallurgical purposes. Owing to the apparent advantages of certain forms of battery, the following may be referred to:

EARTH BATTERY.—The fact that when a piece of copper and a

piece of zinc are imbedded in the earth and connected by a wire, there is a current of electricity obtained from them, in the same way as if they were placed in any battery trough, instantly suggested the application of the earth, or what is termed an earth battery, to the purposes of depositing. We need hardly say these trials were without success. The electricity obtained in this way is very weak, depending wholly upon the moisture of the earth, and the arrangement forming therefore simply a water battery. We have made electrotypes by this means, and also plated small articles, but the action or deposition is very slow. We have obtained a greater amount of deposition in five minutes from one square inch of zinc and copper placed in dilute sulphuric acid, than from four feet of zinc and copper placed in the earth in the space of an hour. An earth battery adapted to deposit from 150 to 200 ounces of silver per day, would require acres of land.

In this as in all other forms of battery the deposit is in relation to the zinc oxidated in the battery; there would therefore be no economy in using the earth battery, and to lessen the amount of surface required by an intensity arrangement, would not alter the law, but rather add to the expense, as it would require upwards of 100 pairs in the earth to be equal to 3 or 4 pairs of Wollaston's for the object of depositing; and would thus be adding to the cost of depositing one hundred times the equivalent of zinc instead of four times its equivalent.

Magneto-Electric Machine.—Several years ago, Mr. Woolrich, of Birmingham, patented a discovery for applying to the deposition of metals the electricity obtained from magnetism or the magneto-electric current, instead of voltaic electricity. We have never had an opportunity of operating with Mr. Woolrich's machine, nor of seeing it in operation for the purpose of deposition. We cannot speak of it from experience; but, from a statement made at the meeting of the British Association in 1850, by Mr. Elkington, of Birmingham, who is the proprietor of the patent, and a gentleman of most extensive experience, it would seem that he had up to that time never been induced to give up the ordinary battery in favor of magnetism or any other suggested improvement. We understand however that this means of obtaining the electricity for the purposes of electro-metallurgy has recently been much improved, by forms of magnets, etc., patented by Mr. Millward, and described by him as follows: "The first branch of the improvement is carried into effect by the employment of an electro-magnet formed by a current of electricity produced from a magneto-electric machine, instead of that generated in a voltaic battery; and such an electro-magnet may be very advantageously used for magnetizing large bars of steel, or for producing very powerful magnets. Any of the known forms of magneto-electric machines will serve thus to convert a bar of steel to an electro-magnet, but the patentee prefers to use one composed of four, eight, or any other number of permanent magnets, having double the number of armatures, and coiled with strong wire of about 60 feet in length. The machine about to be described has been found to answer well in practice. In this machine the steel magnets are composed of eight plates of a U form, weighing about 30 lbs. each plate, and there are eight such compound magnets, all the north poles of which are arranged on one side of the machine, and the south poles on the other side, although this precise arrangement is not essential, and may be varied. The armatures are of soft iron, weighing about 15 lbs., and are coiled about with 60 feet of copper wire of No. 4 gauge, and insulated in the usual manner. The armatures revolve in a brass wheel, and are caused to pass as near to the poles of the magnets as practicable, the commutator or break acting on the whole eight magnets at the same instant, so that the current of electricity shall always pass in one direction, and the surface of the whole of the 64 plates be in combination at the same time. The bar of soft iron used as the electro-magnet with this machine weighs about 500 lbs., and is coiled with bundles of about 30 copper wires of No. 16 gauge, and about 60 feet in length (the bundles are formed by binding a series of uncovered wires together into one covered strand or bundle), and the power of the electro-magnet will depend upon the power of the permanent magnets used in the machine, both as tc the weight it will support from a keeper, and as to its capability of rendering bars of steel permanently magnetic by contact therewith. It will therefore be evident that by having two sets of the permanent magnets, and changing them in such machine, their supporting power may be increased by continued charges or passes from the electro-magnet thus produced. In one form of electro-magnetic machine represented and described under the second head of the invention, the steel bars or permanent magnets are eight in number (these bars may be of cast or soft iron, but when soft iron is employed, bars of steel permanently magnetized will have to be used in conjunction with them), of a U form, and arranged around a circle with their poles pointing towards the centre. Each arm of each of the magnets has attached to it straight bars of steel, also rendered permanently magnetic (of which any desired number, and of any length or size, may be employed, according to the strength of magnet required), which are so placed as to be out of the influence of the armatures when the latter are revolving. The poles of the U-shaped magnets are, on the contrary, as nearly as possible in contact with the armatures which revolve within the circle formed by them, either between the poles or in front of them. Instead of the bars which form the circle being of steel and magnetized, they may be made of soft iron, and depend for their magnetism upon the magnetic bars before named placed around them. In another form of machine both the magnets and armatures are stationary, and the commutator alone has motion between the poles of the horse-shoe magnets and the armatures, being mounted on a spindle and caused to revolve by a band from some driving machinery. The commutator, or break-piece, is composed of a brass centre, with four radial arms of soft iron, either solid or formed of two or more plates."—*See Repertory of Patent Inventions, vol.* 18, *for* 1851, *and Sketches therein.*

The quality of these machines for depositing depends much upon their sustaining power. Eight of these sets of plates or magnets, containing altogether about 12 cwt. of steel, in a proper state of working, are said to form a battery capable of depositing from 12 to 20 ounces of silver per hour, but it is not stated, however, upon what extent of surface this takes place. The relative economy of these magnets over the galvanic battery has not been so great as to recommend their general adoption. We have, however, no doubt that, as such a machine gives electricity of great intensity, it may be superior to the galvanic battery for some purposes, and may give properties to the deposited metals which the ordinary battery does not.*

Before concluding the description of batteries, we may briefly notice one or two little conveniences which are indispensable to the operation. The first of these is what are termed *binding screws,* by which the parts of batteries, as we have shown by numerous figures, are connected together, or by which their poles are connected to the objects through which the voltaic current is to be passed. They are usually made of brass, and of various forms, according to the shape of the objects that are to be connected.

Figures 561 to 566 represent some of the most useful kinds; Fig. 561 is used to connect wires together; Figs. 563 and 565 are required for Smee's battery (see Fig. 559); Fig. 565, to connect the plates of Grove's battery (Fig. 557); Figs. 563 and 564 are used for Daniell's battery; the latter for connection with a zinc rod, the

Figs 561 562 563 564 565 566.

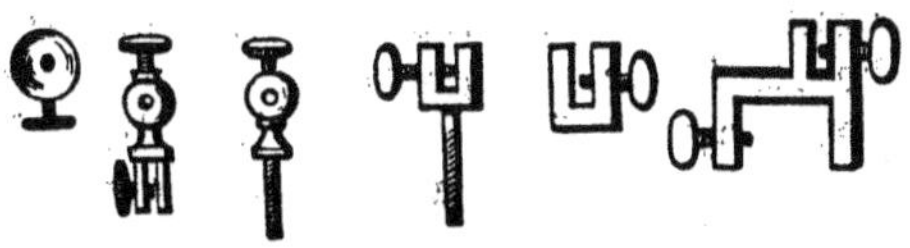

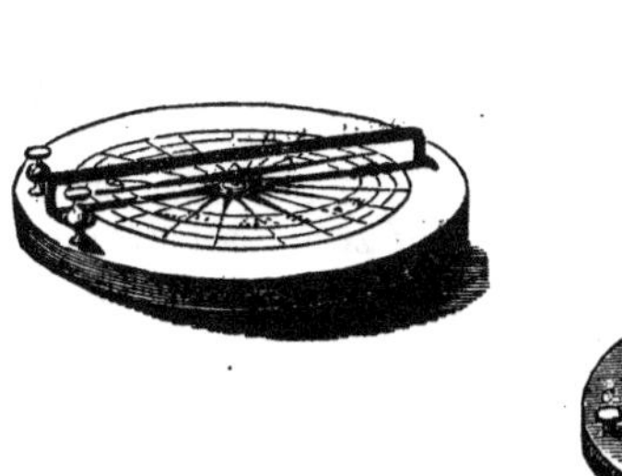

Figs. 567 568.

* For some particulars of the working of electro-magnetic machines, see Shaw's Manual of Electro-Metallurgy, 2d edition, 1843.

former to bind plates together. In all cases, the parts that touch the surfaces to be connected must be perfectly clear and bright.

In many cases, when a complicated apparatus has been put together, it is desirable to ascertain whether the connections are all perfect. This is best determined by means of a galvanometer, two varieties of which are represented by Figs. 567 and 568.

---

# CHAPTER XXVI.

## ELECTROTYPE PROCESSES.

SINGLE-CELL OPERATIONS.—We shall now proceed to detail the process of electrotyping, the materials for which are of the most simple nature. Let us suppose that the object of the student is to copy a copper medal—for example, the side of a penny-piece. Dissolve a quantity of the crystals of sulphate of copper in any convenient vessel; if distilled water can be had, the better. This is conveniently done by suspending the crystals in a coarse cloth on the surface of the water, or the crystals may be put into the water, and well stirred, till dissolved; crushing the crystals facilitates their solution. The water should be kept cold and be fully saturated with the salt, and the solution allowed to stand untouched for several hours. This last precaution is not always essential, but only necessary when the copper solution is not perfectly clear and transparent.

The sulphate of copper of commerce has often a large quantity of iron in it, a portion of which becomes per-oxidized, and will precipitate or fall to the bottom of the solution on standing; indeed, when it is known that the salt contains much iron, it is best to crush the salt very fine, and expose it to the air for some time; when dissolved, after this exposure, a great quantity of iron will settle at the bottom of the solution, which should be carefully decanted and the last portion filtered. The clear solution should now have about one-fourth of its quantity of water added to it, as a completely saturated solution is not the best. A newly-formed solution does not deposit so freely as one that has been in use for some time. The addition of a few drops of sulphuric acid, or, what we have found better, a little sulphate of zinc—about one ounce to the pound of sulphate of copper—improves the condition of a new solution.

Next, put the solution of sulphate of copper into the vessel intended for use, say it is a large jelly-pot, in which let a vessel of unglazed porcelain (porous vessel) be placed, filled to within half an inch of the mouth with a mixture of 24 parts water and 1 sulphuric acid, taking care that the copper solution is of the same depth as the solution in the porous cell.

Preparation of the Coin.—A fine copper wire must now be put round the edge of the coin and fastened by twisting. Then cover the back part, upon which the deposit is not required, with beeswax or tallow, or, what is better, imbed the back of the coin with gutta percha. Have the fore part or face well cleaned, and the surface moistened with sweet oil, by a camel's hair pencil, and then cleaned off by a silk cloth, till the surface appears dry; or, instead of oil, the surface may be brushed over with black lead, which will impart to it a bronze appearance. The use of the oil or black lead is to prevent the deposit adhering to the face of the coin. A very common and excellent method to prevent the copper deposit adhering to the copper mould is this:—Take a gill of rectified spirits of turpentine, and add to it about the size of an ordinary pea of beeswax. When this is dissolved, wet over the surface of the mould with it, and then allow it to dry: the mould is then ready to put into the solution. Medals taken from moulds so prepared retain their beautifully bright color for a long time. But when fine line engravings are to be coated, the little wax dissolved in the turpentine may be objectionable; so also is black lead, for both have a tendency to fill up the fine lines. In this case, let the wash with turpentine be wiped off by a silk handkerchief, instead of drying it: but for ordinary medals this objection will scarcely apply. This being done, the opposite end of the copper wire round the penny-piece is to be connected with a piece of amalgamated zinc, either by means of a binding screw or a hole in the zinc. Then place the zinc in the acid within the porous cell, and put the penny-piece into the copper solution: bring the face of the coin parallel to the zinc, at the distance of about half an inch or one inch from the porous vessel. Deposition immediately begins, and the metal thickens according to the length of time the action is kept up. In about twenty-four hours, the deposit will be of the thickness of a common card, and it may then be taken off. The zinc is to be brushed and washed, before it is put aside. The wire round the coin is now to be untwisted, and by a slight turn will come off easily. The deposit is also easily separated from the mould, which will be a perfect counterpart of the face of the penny-piece.

This mould is next to be treated exactly as described for obtaining it from the penny-piece, and the deposit from it will be a *fac-simile* of one side of the penny-piece. With care, any number of duplicates may be taken from this mould.

It need hardly be remarked, that as copper is deposited the solution becomes proportionally exhausted, and in a short time the current of electricity passing will be too much for the strength of the solution, which will then give a deposit of a sandy consistence, without tenacity.* It is therefore necessary, while the deposition is going on, to suspend some crystals of sulphate of copper at the

* See Laws of Deposition, page 20.

top of the solution, which, as they dissolve, will maintain its strength.

Forms of Apparatus.—It will be observed that no particular form of apparatus is required for electrotyping, but certain modifications may be adopted for convenience and economy. As every portion of the zinc in the acid is capable of giving off electricity, by placing the cell that contains the zinc in the centre of the copper solution, moulds may be suspended on each side of that cell. We have also observed that the zinc plate should not be allowed to touch the cell, as the copper will be reduced upon it and the cell destroyed. To avoid this, the zinc may be suspended by a small wooden peg, put through it and made to rest upon the edges of the cell. Figures 569, 570, 571, 572, represent several convenient forms of apparatus for electrotyping.

Figs. 569 570 571.

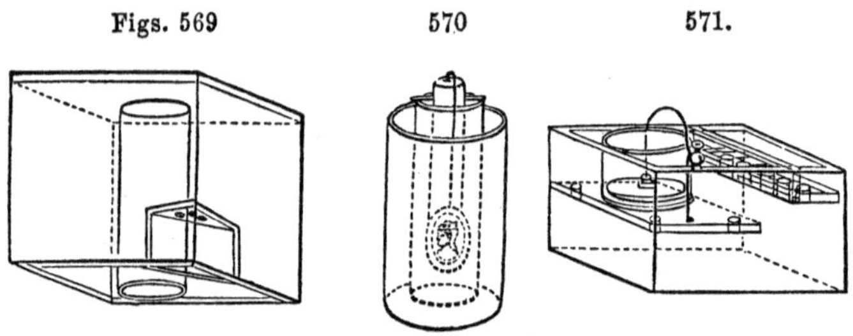

Fig. 572 is a form of the apparatus which the author has used for many years with great success, being both cheap and effective. *a* is a large jelly-pot holding the copper solution; *b* is a flat porous cell, about 4 inches square and half inch wide; *z* the zinc plate. Amalgamated metals may be suspended upon both sides, and the strength of the solution is maintained by suspending a few crystals of the salt in a little cloth bag upon the surface of the solution.

Fig. 572.

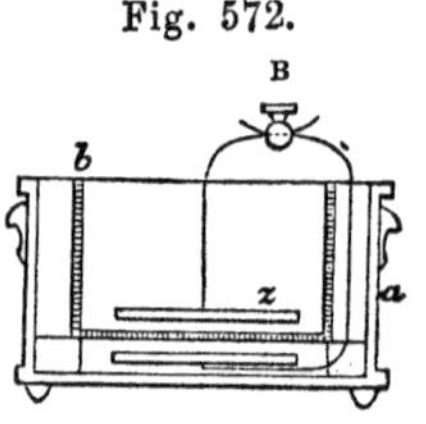


Instead of porous vessels made of earthenware, a bladder may be used, in which the acid and zinc are placed. We have also seen a vessel divided by a porous partition, being either a plate of biscuit porcelain, plaster of Paris, very thin sycamore wood, or dressed skin. The porcelain, as before mentioned, is the best: plaster is too porous, and the solution soon destroys it: wood is too close, and the deposit is consequently very slow: skin does very well for a short time, but it is soon destroyed. When porous cells were not convenient, we have made electrotypes by wrapping the zinc plates in two or three folds of stout cartridge paper, moistened with a solution of salt, and placing this in the copper solution with the mould. Of course, this is only to be adopted when a porous vessel cannot be obtained. The paper lasts but a short time, and has, therefore, to be frequently renewed; besides which, there is always a deposit of copper upon the paper, thus occasioning a loss.

Common coarse garden-pots answer excellently for porous vessels, closing the aperture at bottom by a cork.

The precautions that have been given (see Daniell's Battery, p. 39), as to the preserving of the porous cells when not in use, are applicable to the cells or partitions used in these processes, which, when not in use, should be kept in water, or should not be allowed to dry until they have been in water long enough to dissolve out the salts that were within the pores of the cell; otherwise the salts crystallize, and either crack the cell or cause it to scale off in small pieces. Porous cells, when not thoroughly washed and freed from salts, if laid aside for a few days, often thrown out an efflorescence, or crystalline growth, like mould, of a soft silky texture, and from one-half to one inch in length. An analysis of this efflorescent matter gave

| | |
|---|---|
| Oxide of zinc . . . . | 39·6 |
| Sulphuric acid . . . . | 26·0 |
| Water . . . . . . | 34·2 |
| | 99·8 |

Comparative Value of Exciting Solutions.—We have recommended the porous cell being filled by dilute sulphuric acid, which we consider best; but other saline solutions will serve the same purpose: solutions of common salt, sal ammoniac, sulphate of zinc, have been recommended, and each has been called best in its turn. The following results of experiments with these solutions in the porous cell will show their relative qualities, and enable the student to judge for himself. The size of the zinc plate in the cell used in these experiments measured 6 inches by 6 inches; the copper plates upon which the deposits were formed were the same size; the solution of copper was kept at the same strength; the time that each was in solution was 16 hours.

| Solution in porous Cell. | Zinc dissolved. | | | Copper deposited. | | |
|---|---|---|---|---|---|---|
| | oz. | dwt. | gr. | oz. | dwt. | gr. |
| *Sal ammoniac.* | | | | | | |
| Saturated solution . . . . | 1 | 5 | 14 | 1 | 2 | 12 |
| 1 part saturated solution / 1 part water . . . . | 1 | 8 | 5 | 1 | 3 | 10 |
| 1 part saturated solution / 3 parts water . . . . | 0 | 12 | 13 | 0 | 10 | 17 |
| *Common Salt.* | | | | | | |
| Saturated solution . . . . | 0 | 16 | 3 | 0 | 14 | 12 |
| 1 part saturated solution / 1 part water . . . . | 0 | 18 | 9 | 0 | 17 | 8 |
| 1 part saturated solution / 3 parts water . . . . | 1 | 0 | 5 | 0 | 19 | 16 |
| *Sulphate of Zinc.* | | | | | | |
| Saturated solution . . . . | 1 | 3 | 18 | 0 | 19 | 0 |
| 1 part saturated solution / 1 part water . . . . | 1 | 0 | 8 | 0 | 19 | 18 |
| 1 part saturated solution / 3 parts water . . . . | 0 | 14 | 16 | 0 | 14 | 8 |
| *Sulphuric Acid.* | | | | | | |
| 1 part to 8 of water . . . . | 3 | 1 | 6 | 1 | 4 | 8 |
| 1 part to 16 of water . . . | 2 | 11 | 8 | 2 | 1 | 8 |
| 1 part to 24 of water . . . | 2 | 7 | 3 | 2 | 5 | 6 |

How often Solutions should be Changed and Zinc Amalgamated.—Students have often put this question to us: How often should the solution in the cell be renewed, and the zinc plate be amalgamated? The following are the results of many trials made to ascertain the facts necessary to answer this inquiry. The zinc plates used were nearly one foot square, and the copper plate upon which the deposit was made was of the same size as the zinc plates.

In the first series the zinc plates were not taken out either to brush or re-amalgamate, neither were the solutions renewed during the time specified.

| | | Copper deposited. | | Zinc dissolved. | |
|---|---|---|---|---|---|
| | | oz. | dwt. | oz. | dwt. |
| 2 lb. common salt in one gallon of water | 24 hours . | 12 | 9 | 12 | 17 |
| | 48 hours . | 17 | 13 | 20 | 17 |
| | 60 hours . | 24 | 15 | 34 | 3 |
| 2 lb. sulphate of zinc in one gallon of water | 24 hours . | 9 | 13 | 9 | 18 |
| | 48 hours . | 16 | 4 | 17 | 10 |
| | 60 hours . | 23 | 10 | 24 | 8 |
| 1 lb. sulphuric acid to 24 of water | 24 hours . | 15 | 17 | 17 | 15 |
| | 48 hours . | 27 | 16 | 32 | 3 |

From these results it is evident that the best and most economical manner of treating the solution and the zinc would be to renew the solution every 24 hours, as the second 24 hours do not give, without renewal, above half the deposit of the first 24 hours, while the waste of zinc is very little less than in the first.

The next series of experiments was with the same zinc and the same kind of solutions, but the zinc was taken out every 24 hours, and brushed, but not re-amalgamated, and put back again with new solution in the porous cell.

| | | Copper deposited. | | Zinc dissolved. | |
|---|---|---|---|---|---|
| | | oz. | dwt. | oz. | dwt. |
| Salt and water . | 4 days of 24 hours . | 49 | 16 | 51 | 8 |
| Sulphate of zinc . | 4 days of 24 hours . | 47 | 14 | 48 | 9 |
| Acid and water . | 3 days of 24 hours . | 48 | 13 | 53 | 7 |

These results give the most ample reply to the question so often put, and will guide the manufacturer as well as the student in his operations, whether time or material be of the greatest consequence to him.

We may remark that the sulphate of zinc solution does not require renewal, but simply that we half empty the cell and refill it with water. The sulphate of zinc poured out being nearly saturated, may be crystallized, and will serve for other electro-metallurgical operations.

MAKING OF MOULDS.—The directions giving for obtaining a mould from a penny-piece, by deposition, are applicable to taking moulds from any metallic medal, engraving, or figure that is not undercut; and for depositing within the moulds so produced. On the first discovery of this art, the electrotypist was confined to metallic moulds, as the deposition would not take place except upon metallic surfaces; but the discovery that plumbago, or black lead polished, had a conducting power similar to that of metal, and that the deposit would take place upon its surface with nearly the same facility as upon metal, freed the art at once from many of its trammels, and enabled the operator to deposit upon any substance —wood, plaster of Paris, wax, etc.—by brushing over the surface with black lead. It obliged the electro-metallurgist, however, to render himself expert in the art of moulding, since no good electrotype can be obtained without a perfect mould. We shall, for this reason, endeavor now to give such instructions as will enable the student to make good moulds after a very short practice; but we need hardly add, that in this as well as in every operation, however plain may be the instructions and easy the manipulations, practice is necessary to ensure success; so that the student ought not to lose patience should his first attempt not succeed to his wishes. The substances used for taking moulds from objects to be copied by electrotype are beeswax, stearine, plaster of Paris, and fusible metal; recently, gutta percha has been very successfully used. The articles to be copied are generally composed either of plaster of Paris or metal. Suppose, in the first place, the article to be copied is of metal, and a mould is to be taken from it in wax or stearine. The latter we have not found to answer well alone; when used it should be mixed with wax, about half-and-half.

PREPARATION OF WAX.—Whether the beeswax have stearine in it or not, it is best to prepare it in the following manner:—Put some common virgin wax into an earthenware pot or pipkin, and place it over a slow fire; and when it is all melted, stir into it a little white lead (flake white)—say about one ounce of white lead to the pound of wax; this mixture tends to prevent the mould from cracking in the cooling, and from floating in the solution: the mixture should be re-melted two or three times before using it for the first time.

TO TAKE MOULDS IN WAX.—The medal to be copied must be brushed over with a little sweet oil: a soft brush, called a painter's sash tool, suits this purpose well: care must be taken to brush the oil well into all parts of the medal, after which the superfluous oil must be wiped off with a piece of cotton or cotton wool. If the medal has a bright polished surface, very little oil is required, but if the surface be *matted* or *dead*, it requires more care with the oil. A slip of card-board or tin is now bound round the edge of the medal, the edge of which slip should rise about one-fourth of an inch higher than the highest part on the face of the medal: this done, hold the medal with its rim a little sloping, then pour the wax in the lowest portion, and gently bring it level, so that the melted wax may gradually flow over; this will prevent the formation of air bubbles. Care must be taken not to pour the wax on too hot, as that is one great cause of failure in getting good moulds; it should be poured on just as it is beginning to set in the dish. As soon as the composition poured on the medal is *set* (becomes solid), undo the rim, for if it was allowed to remain on till the wax became perfectly cool, the wax would adhere to it, and being thus prevented from shrinking, which it always does a little, would be liable to crack. Put the medal and wax in a cool place, and in about an hour the two will separate easily. When they adhere, the cause is either that too little oil has been used, or that the wax was poured on too hot.

ROSIN WITH WAX.—Rosin has been recommended as a mixture with wax; mixtures of which, in various proportions, we have used with success; but when often used, decomposition, or some change takes place, which makes the mixture granular and flexible, rendering it less useful for taking moulds. When rosin is used, the mixture, when first melted, should be boiled, or nearly so, and kept at that heat until effervescence ceases; it is then to be poured out upon a flat plate to cool, after which it may be used as de scribed.

MOULDS IN PLASTER.—If a plaster of Paris mould is to be taken from the metallic medal, the preparation of the medal is the same as described above; and when so prepared with the rim of card-board or tin, get a basin with as much water in it as will be sufficient to make a proper sized mould (a very little experience will enable the operator to know this), then take the finest plaster of Paris and sprinkle it into the water, stirring it till the mixture becomes of the consistence of thick cream; then pour a small portion upon the face of the medal, and, with a brush similar to that used for oiling it, gently brush the plaster into every part of the surface, which will prevent the formation of air-bubbles; then pour on the remainder of the plaster till it rises to the edge of the rim: if the plaster is good, it will be ready for taking off in an hour. The mould is then to be placed before a fire, or in an oven, until quite dry, after which it is to be placed, back downwards, in a shallow vessel containing melted wax, not of sufficient depth to flow over the face of the mould, allowing the whole to remain over a slow fire until the wax has penetrated the plaster, and appears upon the face. Having removed it to a cool place to harden, it will soon be ready for electrotyping. If the mould is large and the plaster thick, the wax may be put upon the surface, and only as much as will penetrate a small way into the plaster. In both these instances the wax used is generally lost, and there is always liability of the copper solution passing through, and causing what is termed *surface deposit*, making the face of the medal rough. We may remark that, although occasionally there may be a very good electrotype obtained from a plaster mould, still they are in general very inferior; as the saturating of the plaster has a tendency to blunt the impression, and the wax used for the purpose of saturation becomes expensive. It may be partially recovered by boiling the plaster in water: the wax melts out, and is obtained when the water cools. Plaster should not be used for moulds where wax can be employed, being neither so good nor so economical; but there are cases in which the moulds being very large, the use of plaster is unavoidable.

MOULDS IN FUSIBLE ALLOY.—The next means of taking moulds is by fusible metal. This name is given to alloys of two or more metals which melt at a very low temperature; it suits the purpose of taking moulds of small objects very well. The following are examples of such compositions:

| Tin. | Lead. | Bismuth. | Zinc. |
|---|---|---|---|
| 1 | 1 | 2 | 0 |
| 1 | 2 | 3 | 0 |
| 1 | 0 | 1 | 1 |

These all melt at a temperature below that of boiling water. The ingredients are melted together in an iron ladle, poured out upon a flat stone, broken up, and re-melted in the same way two or three times, in order that they may be thoroughly mixed. The medal from which the mould is to be taken is prepared in the same manner as described for wax.

The fusible alloy is melted and poured into a saucer, or, what does better, a small wooden tray. The operator now watches till it cools down into a semifluid state, or to the point of setting, when he brings the medal suddenly upon it, face downwards, and holds it there until the alloy has fairly set; he then allows it to cool, and undoes the slip around the medal, from which the mould will easily separate. The height of the slip of paper above the surface of the medal determines, of course, the thickness of the mould. The beginner very seldom succeeds in his first attempts at making moulds in fusible alloy; but as a little experience teaches more than the reading of an essay upon the subject, he will soon find both his patience and labor rewarded with gratifying success. Some of the finest moulds are taken by this process, but, from the constant loss of the materials by oxidation, etc., it is expensive, so that its use amongst electro-metallurgists is very limited.

MOULDS IN GUTTA PERCHA.—Gutta percha, as a material for moulding, serves the purpose most admirably. We have seen moulds of this substance equal if not superior to any that we ever saw taken in wax, and of a depth of cutting which it would have been very difficult to have taken in wax. The method adopted for taking moulds is to heat the gutta percha in boiling water, or in a chamber heated to the temperature of boiling water, which makes it soft and pliable. The medal is fitted with a metallic rim, or placed in the bottom of a metal saucer with a cylindrical rim a little larger than the medal; the medal being placed back down, a quantity of gutta percha is pressed into the saucer, and as much added as will cause it to stand above the edge of the rim. It is now placed in a common copying-press and kept under pressure until it is quite cold and hard. The impressions taken in this way are generally very fine. When the medal is not deep cut a less pressure may suffice, but when the pressure is too little the impression will be blunt.

Gutta percha takes a coating of black lead readily, and the deposit goes over it easily.

A mixture of gutta percha and marine glue has been recommended for moulds as superior to gutta percha alone. We have not had an opportunity of using this mixture, but have every confidence in the recommendations given of it.

MOULDS FROM FERNS, SEA-WEED, ETC.—A method of taking impressions of fern-leaves and sea-weeds has recently been proposed by Dr. F. Branson, in the *Athenæum*. It is thus described:

"A piece of gutta percha, free from blemish, and the size of the plate required, is placed in boiling water. When thoroughly softened it is taken out and laid flat upon a smooth metal plate and immediately dusted over with the finest bronze powder used for printing gold letters. The object of this is threefold—to dry the surface, to render the surface more smooth, and to prevent adhesion. The plant is then to be neatly laid out upon the bronze surface and covered with a polished metal plate either of copper or of German silver. The whole is then to be subjected to an amount of pressure sufficient to imbed the upper plate in the gutta percha. When the gutta percha is cold, the metal plate may be removed and the fern gently withdrawn from its bed. A beautiful impression of the fern will remain." An electrotype may be deposited upon the bronzed or black leaded gutta percha.

We have seen many electrotype leaves done by this method, which were certainly very pretty as electrotypes, and the process is well adapted for flat leaves; but the pressure required renders it unsuitable for some kinds of leaves—indeed, it destroys the natural forms of the greater number both of leaves and sea-weeds. The products of the process cannot, indeed, be compared with those electrotype leaves the moulds of which are taken by wax. The great merit of the process is its ease and simplicity. The method given for taking the mould of the leaf is suitable for any kind of flat mould in gutta percha. The mould of a leaf may be taken in plaster by placing the leaf upon dry sand and pressing the sand under and on each side to fill up the spaces under the leaf so as to bear the pressure of the plaster, putting a collar of paper round the sand to prevent its yielding, and then pouring the plaster over the whole. When the plaster is set, the leaf is removed and the plaster trimmed round with a knife. This also has its difficulties; for when leaves have hairs upon them they stick into the plaster. The method of taking moulds of leaves in wax is by holding the leaf in the hand and brushing a thin layer of melted wax over the surface to be moulded; allowing this to harden, and then brushing on another layer—and so on until the wax is sufficiently thick to suffer handling. The leaf is then gently drawn off the wax, which is to be black-leaded, and put into the electrotype apparatus to receive the coating of copper. A type of the leaf is by this means obtained with all its natural convolutions.

Nature Printing.—A further improvement upon making moulds of leaves and other vegetable objects, has been practised by an eminent firm in London. The leaf is carefully dried and laid upon a smooth piece of milled lead, which is placed between two steel plates, and passed between rollers, these press the leaf into the lead, and produce a complete mould. Copies from this may be taken with gutta percha or electrotype. Printed impressions of leaves, sea weed, and such like objects, prepared in this way, may be seen in an excellent work published by Bradbury and Evans, and a full detail of the process, with specimens, will be found in the proceedings of the Royal Institution for 1854.

Casting of Reptiles, etc.—Imbed the subject in a mould made of four parts of plaster of Paris, one of unburnt lime powder, and one of Flanders brick-dust. Dry the mould carefully, then make it red hot, and burn the subject out of it, taking care to free the mould from the ashes. Fusible metal may be cast in this mould, and then be covered with copper, from which the alloy may be afterwards melted, or a wax model may be taken of the object, pouring the wax in just before setting. In neither case must the mould be melted until after the model, whether of alloy or wax, is taken, when the whole is placed in water, the lime causes the mould to dissolve or break up, and the figure modelled within it may be taken and covered with copper. Flowers, insects, lizards, and other little animals may be typed in this way. In all these processes, perseverance and care are the best cures for little difficulties.

WAX MOULDS FROM PLASTER.—If the object, which we assume to be a medal, from which the mould is to be taken, be composed of plaster of Paris, and the mould to be taken is in wax, the first operation is to prepare the plaster medal. Some boiled linseed-oil, such as is used by house painters, is to be laid over the surface of the medal with a camel's hair pencil, and continued until it is perfectly saturated, which is known by the plaster ceasing to absorb any more of the oil. This operation succeeds best when the medal is heated a little. The medal should now be laid aside till the oil completely dries, when the plaster will be found to be quite hard, and having the appearance of polished marble; it is, consequently, fit to be used for taking the wax mould, which is done in the same manner as we have described for taking a wax mould from a metallic medal.

Many prefer saturating the medal with water: this is best done by placing the medal back down in the water, but not allowing it to flow over the face; the water rises, by capillary attraction, to the surface of the medal, rendering the face damp without being wet. The rim being now tied on the plaster medal, the melted wax is poured upon it. This method is equally good, but liability to failures is much greater, caused generally by the wax being too hot.

The plaster medal may also be saturated with skimmed milk and then dried; by repeating this twice, the plaster assumes on the surface an appearance like marble, and may be used for taking wax moulds.

MOULD OF PLASTER FROM PLASTER MODELS.—When a plaster mould is to be taken, the face of the model is prepared differently to that described, in order to prevent the adhesion of the two plasters. The best substance we have tried for this purpose is a mixture of soft soap and tallow, universally used by potters for preparing their moulds, and called by them *lacquer*. It is prepared in the following manner:—half-a-pound of soft soap is put into three pints of clean water, which are set on a clear fire, and kept in agitation by stirring; when the mixture begins to boil, add from one ounce to an ounce and a-half of tallow, and keep boiling till it is reduced in bulk to about two pints, when it is ready for use. The surface of the medal must be washed over with this lacquer, allowing it to absorb as much as it can, when it assumes the appearance of polished marble; it is now prepared with a rim of paper, and the mould taken as directed for taking plaster moulds from metallic medals. When hardened, they will separate easily.—Wetting the plaster model with a solution of soap before taking the cast will do, or, if the plaster model has been saturated with oil or milk, it has only to be moistened with sweet oil the same as a metal model.

FUSIBLE ALLOY FROM PLASTER.—If a mould of fusible metal be required from a plaster medal, the plaster may be saturated either with boiling oil or the soap and tallow lacquer, and the mould taken in the same manner as from a metallic medal.

COPPER MOULDS FROM PLASTER.—Many electro-metallurgists prefer taking a mould in copper when the medal is of plaster of Paris. This is done by the electrotype process; the plaster model is saturated with wax over a slow fire, as already detailed, and then prepared for taking an electrotype in the usual manner (see page 534). We need hardly mention that the model in this case is destroyed; but, notwithstanding, in the case of plaster models, to take a copper mould is the most preferable, as it may be repaired in case of slight defect, and it may be used over and over again without deterioration.

When an electrotype is required of a model that is undercut, or of a bust or figure, the process which we have described will not answer, as the mould cannot separate from the model. In such circumstances the general method of proceeding is to part the mould in separate pieces, and then join these together. The material used for this purpose is plaster of Paris. The operation, however, to be well done, requires a person of considerable experience.

ELASTIC MOULDING.—The process patented by Mr. Parks for taking a mould of any kind of model in one piece, is excellently adapted for the electrotypist. The material is composed of glue and treacle. Twelve pounds of glue is steeped for several hours in as much water as will moisten it thoroughly; this is put into a metallic vessel, which is placed in boiling water as a hot bath. When the glue falls into a fluid state, three pounds of treacle are added, and the whole is well mixed by stirring. Suppose, now, that the mould of a small bust is wanted, a cylindrical vessel is chosen so deep that the bust may stand in it an inch or so under the edge. The inside of this vessel is oiled, a piece of stout paper is pasted on the bottom of the bust to prevent the fluid mixture from going inside, and if it is composed of plaster, sand is put inside to prevent it from swimming. It is next completely drenched in oil and placed upright in the vessel. This done, the melted mixture of glue and treacle is poured in till the bust is covered to the depth of an inch. The whole must stand for at least twenty-four hours, till it is perfectly cool throughout—after which it is taken out by inverting the vessel upon a table, when, of course, the bottom of the bust is presented bare. The mould is now cut by means of a sharp knife, from the bottom up the back of the bust to the front of the head. It is next held open by the operator, when an assistant lifts out the bust and the mould is allowed to reclose. A piece of brown paper is tied round it to keep it firm. The operator has now a complete mould of the bust in one piece; but he cannot treat it like wax moulds, as its substance is soluble in water, and would be destroyed if put into the solution. A mixture of wax and rosin, with occasionally a little suet, is melted and allowed to stand till it is on the point of setting, when it is poured carefully into the mould and left to cool. The mould is then untied and opened up as before; the wax bust is taken out; and the mould may be tied up for other casts. Besides wax and rosin there are several other mixtures used—deer's fat is preferable to common suet, stearine, etc. The object is to get a mixture that takes a good cast and becomes solid at a heat less than that which would melt the mould.

Moulding of Figures.—If the model or figure be composed of plaster of Paris, a mould is often taken in copper by deposition. The figure is saturated with wax, as described for a medal, and copper deposited upon it sufficiently thick to bear handling without damage when taken from the model. The figure with the copper deposit is carefully sawn in two, and then boiled in water, by which the plaster is softened and easily separated from the copper, which now serves as the mould in which the deposit is to be made. It is prepared in the same way as we have described for depositing in copper moulds. When the deposit is made sufficiently thick, the copper mould is peeled off and the two halves of the figure soldered together. The copper moulds which are deposited upon the wax models taken in the elastic moulding are often treated in the same manner; but more generally these moulds are used for depositing silver or gold into them, to obtain *fac-similes* of the object in these metals, in which case the copper moulds are dissolved off by acids, as will be described in a subsequent section.

Figures covered with Copper.—When plaster busts or figures are wanted in copper, the most usual way is to prepare the figure with wax as described, and to coat it over with a thin deposit of copper, letting the copper remain. Some operators, when it can be done, remove the plaster and wash over the inside with an alloy of tin and lead melted. In this case the copper must previously be cleaned by washing first in a solution of potash, and then with chloride of zinc. The latter mode will cause the alloy to adhere to the copper and give it strength. In either of these cases the deposit must not be very thick, or it will throw the figures out of proportion, such as the features of a bust, etc. Any slight roughness of deposit may be easily smoothed down by means of fine emery.

The Preparation of Non-Metallic Moulds to receive Deposit.—Having detailed what we have found best for obtaining moulds of objects for the purpose of electrotyping, we proceed to the manner of obtaining a deposit upon these moulds. Were any of the plaster or wax moulds attached to the zinc and immersed in the copper solution in the same manner as described with the penny-piece (page 534), no deposit would be obtained, because neither the plaster nor the wax is a conductor of electricity. Some substance must now be applied to the surface in order to give it conducting power. There are several ways of communicating this property, but the best and most simple for the articles under consideration is to apply common black lead (already referred to) in the following manner:—A copper wire is put round the edge of the medal, or, if wax moulds are used, a thin slip of copper may be inserted into the edge of the mould—or, being slightly heated and laid upon the back, the two will adhere. A fine brush is now taken (we have found a small hat-brush very suitable) and dipped into fine black lead, and brushed over the surface of the medal. The brushing is to be continued until all the face round to the wire upon the edge, or slip of copper forming connection, has a complete metallic lustre. A bright polish is necessary to the obtaining a quick and good deposit.

In brushing on the black lead, care should be taken not to allow any to go upon the back or beyond the copper connection, or the deposit will follow it, and so cause a loss of copper, and make the mould more difficult to separate from the deposit; being, as it were, incased. If the electrotypist takes the labor himself of filing off all the superfluous copper from the edge of his deposited medal, it will do more than any written precautions to teach the necessity of preventing as much as possible the deposit going further than is necessary. When the face of the mould is properly black-leaded, the copper wire connected with it is attached to the zinc plate in the porous cell, and the mould immersed in the copper solution; the deposit will immediately begin upon the copper connection, and will soon spread over every part, covering the black-lead polish with less or more facility according to the state of the solutions and other circumstances to be afterwards noticed. When the deposit is considered sufficiently thick for removing—which, in ordinary circumstances, will require from two to three days—the medal is taken out of the solution, and washed in cold water, and the connection is taken off. If the deposit has not gone far over the edge of the mould, the two may be separated by a gentle pull; if otherwise, the superfluous deposit must be eased off, and if care be taken the wax may be fit to use over again: but when the mould is plaster of Paris, however well it may be saturated with wax, it is seldom in a condition to use again. If the plaster mould be large and thick, it is advisable to coat the back with wax or tallow, which is done by brushing it over with either substance in a melted state: the mould being cold will not absorb the wax or tallow: hence it may be recovered again. The sulphate of copper possesses so penetrating a quality that if the slightest imperfection occurs in the saturation of the mould by wax, the solution will penetrate through it, and the copper will be deposited upon the face of the object adhering to the plaster, giving to the metal a rough, matted appearance, and seriously injuring it.

USING METAL MOULDS.—The mould in fusible alloy does not require to be black-leaded, but the back and edge must be protected by a coating of wax or other non-conducting material; it may be connected in the same way as the penny-piece (page 534) by putting a wire round its edge previous to laying on the non-conducting substance, such as tallow or wax, which should also cover the wire. Or a slip of copper, or wire, may be laid upon the back and fastened by a drop or two of sealing-wax; the back is then coated: but care must be taken that the wax do not get between the connection and the medal which will prevent deposit. The deposit on this mould goes on instantaneously, the same as over the penny-piece. When sufficiently thick, it may be taken off in the same manner as from the wax mould, the surface having been prepared by turpentine (page 534) to prevent adherence. These moulds may be used several times, if care be taken not to heat them, as they easily melt.

The medals obtained from metallic moulds prepared with the turpentine solution have a bright surface, which is not liable to change easily, but if the mould has been prepared with oil or composed of wax or plaster, the metal will either be dark, or will very easily tarnish. The means of preserving them, either by bronzing or plating with other metals, will be detailed in a subsequent section.

PRECAUTIONS ON PUTTING THE MOULDS INTO A SOLUTION.—In putting moulds into the copper solution, the operator is often annoyed by small globules of air adhering to the surface, which either prevent the deposit taking place upon these parts, or, when they are very minute, permit the deposit to grow over them—causing small hollows in the mould, which give a very ugly appearance to the face of the medal. To obviate this, give the mould, when newly put into the solution, two or three shakes, or give the wire attached to it, while the mould is in the solution, a smart tap with a key or knife, or any thing convenient; but the most certain means we have tried, is to moisten the surface with alcohol just previous to putting it into the copper solution. A little practice in these manipulations will soon enable the student to avoid these annoyances.

DEPOSITION ON LARGE OBJECTS.—When busts or figures, whether of wax or plaster of Paris, are to be coated with copper, with no other conducting surface than black lead, it is attended with considerable difficulty to the inexperienced electrotypist. The deposit grows over all the prominent parts, leaving hollow places, such as armpits, neck, etc., without any deposit; and when once missed, it requires considerable management to get these parts coated, as the coated parts give a sufficient passage for the current of electricity. It is recommended by some electrotypists to take out the bust, and coat the parts deposited upon with wax, to prevent any further deposit on them; but this practice is not good, especially with plaster of Paris, for an electrotype ought never to be taken out till finished. Sometimes the resistance of the hollow parts is occasioned by the solution becoming exhausted from its position in regard to the positive pole. In this case a change of position effects a remedy. It may be remarked that when a bust or any large surface having hollow parts upon it, is to be electrotyped, as many copper connections as possible ought to be made between these parts and the zinc of the battery. Let the connections with the hollow parts be made with the finest wire which can be had, and let the zinc plate in the cell have a large surface compared to the surface of the figure, and the battery be of considerable intensity; if attention is paid to these conditions, the most intricate figures and busts may be covered over in a few hours. Care has to be observed in taking off the connections from the deposit, or the operator may tear off a portion of the deposit: if the wires used are fine, they should be cut off close to the deposited surface.

TO MAKE BUSTS AND FIGURES.—Busts and figures, and other complicated works of art, which cannot be perfectly coated with black lead, may be covered by a film of silver or gold, which serves as a conducting medium to the copper. This is effected by a solution of phosphorus in sulphuret of carbon. The operation being patented, we will take advantage of the description given of it in the specification. "The solution of phosphorus is prepared by adding to each pound of that substance 15 lbs. of the bisulphuret or other sulphuret of carbon, and then thoroughly agitating the mixture; this solution is applicable to various uses, and amongst others, to obtaining deposits of metal upon non-metallic substances, either by combining it with the substances on which it is to be deposited, as in the case of wax, or by coating the surface thereof. Any of the known preparations of wax, may be treated in this way, but the one preferred is composed of from 6 to 8 ounces of the solution: 5 lbs. of wax, and 5 lbs. of deer's fat, melted together at a low heat, on account of the inflammable nature of the phosphorus. The article formed by this composition is acted upon by a solution of silver or gold in the manner hereinafter described with respect to articles which have been coated with the solution."*

COATING OF FLOWERS, ETC.—"If the solution is to be applied to the surface of the article, an addition is made to it of one pound of wax or tallow, one pint of spirits of turpentine, and two ounces of India rubber, dissolved with one pound of asphalt, in bisulphuret of carbon, for every pound of phosphorus contained in the solution. The wax and tallow being first melted, the solution of India rubber and asphalt is stirred in; then the turpentine, and after that the solution of phosphorus is added. The solution prepared in this manner is applied to the surfaces of non-metallic substances, such as wood, flowers, etc., by immersion or brushing;

* Repertory of Patent Inventions, 1844.

the article is then immersed in a dilute solution of nitrate of silver, or chloride of gold; in a few minutes the surface is covered with a fine film of metal, sufficient to insure a deposit of any required thickness on the article being connected with any of the electrical apparatus at present employed for coating articles with metal. The solution intended to be used is prepared by dissolving four ounces of silver in nitric acid, and afterwards diluting the same with twelve gallons of water; the gold solution is formed by dissolving one ounce of gold in nitro-muriatic acid, and then diluting it with ten gallons of water."

We have frequently repeated the operations described by this patentee with entire satisfaction, and were enabled to cover every variety of surface with great facility.

The solutions of silver and gold, prepared as above, will last for a long time, and do a great many articles. When it is convenient it is best to use both solutions. The connecting wire should first be attached to the article to be coated, before being dipped into the phosphorus solution, but connected at such parts as will not hurt the appearance of the object by leaving a mark when it is taken off. Care should be taken not to touch the article with the hands after it is dipped into the solution. The object supported by the connections is immersed in the phosphorus solution, where it remains for two or three minutes. When taken out it is dipped into the silver solution, and as soon as the surface becomes black, having the appearance of a piece of black china, it is to be dipped several times in distilled water, and then immersed in the solution of protochloride of gold about three minutes: the surface takes a bronze tinge by the reduction of the gold. It is next washed in distilled water by merely dipping, not by throwing water upon it. The wire connection is now attached to the zinc of the battery, and then the article put into the copper solution, and in a few minutes the article is coated over with a deposit of copper. A thin copper surface may thus be given to small busts or figures without sensibly distorting the features by want of proportion.

Figures from Elastic Moulds.—When taking a wax cast from the elastic mould, described in page 544, we prefer the phosphorized mixture. After taking out the mould it is only necessary to make the connections, and pass it through the gold and silver solutions, as described, and then to connect it with the battery.

We may also mention that the principal object of making copper moulds by this process, in the manufactory, is not to make *facsimiles* in copper, but to make articles of solid silver or gold. Copies of highly wrought work, either chased or engraved, or of articles, duplicates of which cannot be obtained, or of which the workmanship is costly, may by this means be made in solid silver or gold, at little more expense than the cost of the metal. Having obtained the copper mould, silver is deposited in it to any thickness, and the copper dissolved off. However, an extensive trade is now being carried on in figures and other works of art deposited in copper and then bronzed, which gives them an appearance often not much inferior to that of antique works of the highest art.

ELECTROTYPES FROM DAGUERREOTYPES.—What may be justly termed the perfection of electrotyping, is the production of electrotypes from daguerreotypes. The daguerreotype picture being taken, a small portion of the back is cleaned with sand-paper, taking care not to allow any thing to touch the face; a little fine solder is placed on this part; a piece of flattened wire, also cleaned, is placed upon the solder, the whole moistened with dilute muriatic acid, or chloride of zinc. The wire is now held over the gas or a lamp about half an inch from the plate; the heat is transmitted through the wire to the solder, which melts, and the wire is soldered to the type; the back is then protected by wax, and the daguerreotype is now put into the copper solution in the same manner as a medal; the deposit proceeds rapidly, and when sufficiently thick the two easily separate, and an impression of the picture is obtained from the daguerreotype with an expression softer and finer than the original; several electrotypes may, with care, be taken from one picture. The electrotype may now be passed through a weak solution of cyanide of gold and potassium, in connection with a small battery, and thus a beautiful golden tint be given to the picture, which serves to protect it from the action of the atmosphere; but they should also be protected by a glass, which may be fixed on in the manner pointed out in another section. The most successful operators that we have known in this and every other department of electrotyping are Dr. Thomas Paterson, of Glasgow, and Mr. Bawtree, of London.

DEPOSITING BY SEPARATE BATTERY.—Having described, so far as we know them, the best and most simple means of obtaining moulds, and their preparation for receiving the deposit of the metal, we return again to the management of solutions and batteries, and the application to other metals besides copper.

Although in our account of the porous or single cell system (page 533) we have recommended it as the best and most economical for electrotyping, still many eminent electro-metallurgists prefer using the battery system; and indeed there are solutions of copper and of other metals to which the porous cell system cannot be applied, from the nature of the solution and the necessity of intensity to decompose them.

While depositing upon a mould by the single cell, let the wire which connects them be cut in the middle, and a mould be attached to the end of the portion remaining upon the zinc plate, and a small plate of copper to the end of the wire remaining upon the mould in the copper solution, and let these two be put into a second vessel containing a solution of sulphate of copper. The action between the zinc and medal in the double or first cell will go on as before—namely, the electricity passing through the porous cell and the solution to the medal; but on returning to the zinc it must pass through the copper solution, which is in the second vessel, between the mould and copper plate, where it produces the same effects as in the first cell. The sulphuric acid is liberated at the copper plate and dissolves it, and the copper is deposited upon the mould, so that the solution in this cell is maintained at one strength; hence there is no necessity for hanging crystals of sulphate of copper in this solution.

It will be observed that the electricity having to pass through a second solution, is made to perform double duty, and must consequently be much more economical. We found the results to be these: A single cell, with a mould, was placed two inches from the porous cell, and of the same size as the zinc plate; and another, similarly arranged, but connected with a metal mould and copper plate of similar size to the zinc and copper, was placed one inch apart in the copper solution of second cell. The mould in the single cell had gained 100 grains, and the zinc plate lost 108 grains. The mould in the battery cell of the other arrangement had only deposited upon it 30 grains—the zinc plate had lost 35 grains; but the mould in the second or decomposition cell had also deposited upon it 30 grains, making in all 60 grains deposited for 35 zinc dissolved, but taking nearly double the time. These arrangements, as we have before observed, are simply a modification of a single pair of Daniell's battery connected with a decomposition cell, the advantages of which are not applicable to any other battery, as in no other battery does deposition take place within the battery cells: indeed this method of using a compound depositing apparatus is very seldom employed. Batteries of a different form, as Smee's are generally adopted. Figure 573 represents a Smee's battery, connected with a medal and copper plate in a separate or decomposition cell; and Fig. 574 is a large decomposition trough for doing several medals at one time. Of course any battery may be attached to these medals and plate by the brass connections seen on the end of the trough. Bear in remembrance

Fig. 573.

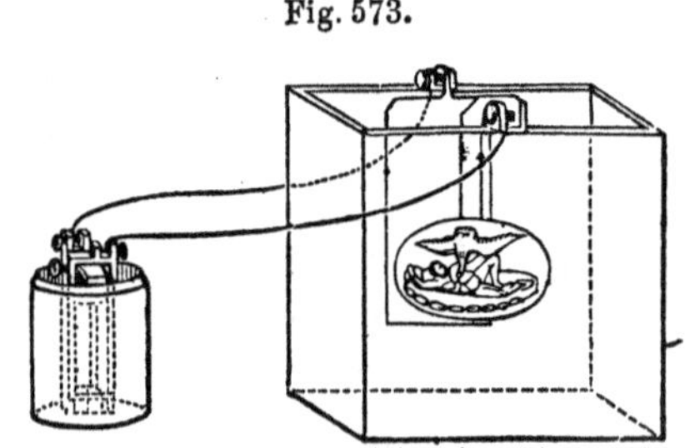

Fig. 574.

that the zinc of the battery is connected with the medals, and the copper or platinized silver with the copper in the decomposition cell.

SIZE OF THE ELECTRODES.—When a separate battery is used for the purpose of depositing in a decomposition cell, there are several conditions which are well to be observed, as they influence the amount and character of the deposit. The first is the size of the electrodes or medals, in relation to the zinc in the battery. The results we have obtained may be expressed in general terms. When the deposit upon electrodes of the same size as the zinc plates in a Wollaston's battery equals 100, that upon electrodes one half the size of the zinc plates in the battery will be equal to 57, and upon electrodes double the size of the zincs of the battery 190; but this last condition is affected by the intensity of the arrangement.

RELATIVE POWER OF BATTERIES.—The following experiments, made with electrodes double the size of the zinc plates of the batteries, all at equal distances (1 inch apart), will show the relative power of the batteries. The time in action was one hour each: only one pair of plates constituted the battery.

| | | |
|---|---|---|
| Grove's battery deposited . . | 104 | grains. |
| Single cell . . . . . . . . | 62 | " |
| Daniell's . . . . . . . . . | 33 | " |
| Smee's . . . . . . . . . | 22 | " |
| Wollaston's. . . . . . . . | 18 | " |

CONSTANCY OF BATTERIES.—But the first hour of the action of most batteries differs from an hour afterwards, so that one kind of battery may be most useful for a short time, and another sort if the action is to be continued for a length of time. The following table will illustrate this remark, the condition being the same as in last experiment, or the last experiments being continued, and the results taken every hour for seven successive hours:

| | 1 hour. | 2 hours. | 3 hours. | 4 hours. | 5 hours. | 6 hours. | 7 hours. | Total. |
|---|---|---|---|---|---|---|---|---|
| Grove's battery . | 104 | 86 | 66 | 60 | 54 | 49 | 45 | 464 grs. |
| Single cell . . . | 62 | 57 | 54 | 46 | 39 | 29 | 24 | 311 " |
| Daniell's . . . | 33 | 35 | 34 | 32 | 32 | 30 | 31 | 227 " |
| Smee's . . . . | 22 | 16 | 14 | 11 | 12 | 11 | 10 | 96 " |
| Wollaston's . . | 18 | 14 | 15 | 12 | 11 | 10 | 10 | 90 " |

To make this comparision more practical, larger plates were used for the battery, and proportionately larger electrodes, and the battery kept in operation until one pound of copper was deposited, renewing the acid, and brushing the zincs every 24 hours. The time taken to effect this was:

| | |
|---|---|
| Grove's Battery | 19½ hours. |
| Single cell | 45 " |
| Daniell's* | 49 " |
| Smee's | 147 " |
| Wollaston's | 151 " |

COMPARATIVE PRODUCE OF BATTERIES.—The expense of the materials used in these experiments was as follows (of course the materials will differ in cost both at different times and in different localities, and more common materials may be used):

By the process with Grove's battery, one pound of deposited copper costs

| | |
|---|---|
| 1 lb. Copper, from positive electrodes | 25 cts. |
| 1¼ lb. Amalgamated zinc | 21 " |
| 1½ lb. Nitric acid | 19 " |
| Sulphuric acid | 02 " |
| | 67 cts. |
| Add time, say one cent per hour, for comparison | 20 " |
| | 87 cts. |

By single cell apparatus, one pound of deposited copper costs

| | |
|---|---|
| 1 lb. Sulphuric acid | 04 cts. |
| 1$\frac{1}{16}$ lb. Amalgamated zinc | 17 " |
| 4 lb. Sulphate of copper | 37 " |
| | 58 cts. |
| Time, at one cent per hour | 45 " |
| | $1.03 |

By Daniell's battery, one pound of deposited copper costs

| | |
|---|---|
| 1$\frac{2}{16}$ lb. Amalgamated zinc | 19 cts. |
| 4 lb. Sulphate of copper | 37 " |
| 1 lb. Copper from electrode | 25 " |
| Sulphuric acid | 02 " |
| | 83 cts. |
| Time, at one cent per hour | 49 " |
| | $1.32 |

* The Daniell's battery used in this experiment had flat plates, not circular, as described at page 523.

By Smee's battery, one pound of deposited copper costs

| | |
|---|---|
| $1\frac{1}{2}$ lb. Amalgamated zinc | 21 cts. |
| 3 lb. Sulphuric acid | 12 " |
| 1 lb. Copper from electrode | 25 " |
| | 58 cts. |
| Time, at one cent per hour | $1.47 |
| | $2.05 |

By Wollaston's battery, one pound of deposited copper costs

| | |
|---|---|
| 1 lb. Copper from electrode | 25 cts. |
| $1\frac{2}{16}$ lb. Amalgamated zinc | 19 " |
| 3 lb. Sulphuric acid | 12 " |
| | 55 cts. |
| Time, at one cent per hour | $1.51 |
| | $2.06 |

By thus adding the time, at a given rate, it serves to illustrate what we have before stated respecting the necessity of placing the value of time against the cost of materials. In manufactories, where time has to be paid for, it may be cheapest to use the battery with the most costly materials; but where time is of no consideration, or, as is often the case, if, while the operations are going on, the workmen are employed in other necessary labor, a cheaper apparatus will answer: but the student or manufacturer will, by the above general results, be enabled to choose the process most suitable for his purposes. It must be borne in mind that an allowance has to be made on the first, second, and third, for wear and tear of the porous vessels, not included in the above estimate. Although the results of these experiments give, exclusive of time, the cost of one pound of electrotyped copper—thus

| | |
|---|---|
| Grove's battery | 67 cts. |
| Single cell | 58 " |
| Daniell's | 83 " |
| Smee's | 58 " |
| Wollaston's | 55 " |

still we know from long experience in the use of single cell, Smee's, and Wollaston's batteries, for manufacturing purposes, that the price of the pound of copper deposited may be more correctly stated at 62 cents—there being always loss in making the purest article (the copper) from impure materials, as the sulphate of copper, or the ordinary copper of commerce which is used as electrodes.

Mr. Smee, in his "Advice to capitalists who propose entering upon the business of electro-metallurgy," gives a table of expenses incurred by the use of different batteries. But his rules are based too exclusively upon theoretical considerations, and without that regard for practical conditions which are so important to the manufacturer. Mr. Smee recommends for use what he calls "an odds-and-ends battery," composed of odd scraps of zinc put into acid, having in the same vessel a piece of copper or platinized silver and a wire placed in contact with them which forms the electrode. This battery may be convenient for the amateur electrotypist, as it enables him to use up all his waste zinc. Raw zinc, or spelter, Mr. Smee says, may also be used in this way, constituting the cheapest of all batteries for manufacturing purposes. The data of his calculations are as follows:—The copper sheet forming the positive electrode is quoted at 25 cents per lb.; wrought zinc, 14 cents per lb.; raw zinc at a little more than half the price of wrought zinc, which we will call 8 cents per lb., although he rates it at 10 cents. Iron is given at from 2 to 4 cents per lb. The equivalent weight of copper is given at 32, of zinc at 32, and of iron at 28; that is to say, 32 parts (say ounces) of zinc dissolved in a battery will, or should deposit 32 ounces of copper; and if iron be used, 28 ounces of iron should deposit 32 ounces of copper. Hence, in the plain language of a manufacturer, we should say that, with an odds-and-ends battery and raw zinc, there would be, for every pound of copper deposited,

| | |
|---|---|
| 16 oz. of zinc used . . . . | 08 cts. |
| 16 oz. of copper dissolved . . | 25 " |
| | 33 cts. |

And when iron is used, the expense of depositing 1 lb of copper would be,

| | |
|---|---|
| 16 oz. of iron, *say* . . . . | 04 cts. |
| 16 oz. of copper . . . . . | 25 " |
| | 29 cts. |

Notwithstanding these results, Mr. Smee proves, by several fractional formulæ and an algebraic equation, that the cost of depositing a pound of copper is

| | |
|---|---|
| By iron . . . . . . . . | 37 cts. |
| By odds-and-ends battery . . | 25 " * |
| | 62 cts. |

To this it may be replied by the manufacturer, that, in the first place, raw zinc or spelter used in the way described for an odds-and-ends battery would lose two or three times the quantity that is stated for every equivalent of copper; and secondly, that this form

* Smee's Elements of Electro-metallurgy, 3d edition, p. 112.

of battery is altogether unsuitable for manufacturing purposes, even when amalgamated scrap zincs are used; and, as regards the calculation, it is not easy to see that, while a pound of copper, dissolved from the positive electrode, originally costs 25 cents, it could, notwithstanding, be deposited by the destruction of 1 lb. of zinc, not including acid, etc., at the expense of only 25 cents. It ought always to be remembered, that, for manufacturing purposes, the surface upon which the metal is to be deposited in general amounts to several square feet. The article may be, for example, a large ornamental vase, having four square feet of surface. An odds-and-ends battery, or an iron single pair battery would be too weak. To deposit, with a separate battery, upon a surface such as that of the vase, it requires two or three pairs of plates to give what we may call economical power.

RECOVERY OF MERCURY FROM WASTE ZINC.—The general practice of manufacturers, when the scraps of zinc become small, is either to treat them as referred to at page 512, to distil the mercury from the zinc, or to sell the scraps to parties who do distil them. This is done by putting the scraps into an iron retort, subjecting it to a red heat, and allowing the beak of the retort to pass into a condenser, which has a tube dipping into water. The mercury distils over, and condenses in the water. The zinc left in the retort is found to be so impure as not to be fit to melt and roll again, but it may be used in the composition of common brass. Mr. De la Rue, in a communication to the Chemical Society,* gives the results of several analytical experiments upon scrap zinc. Before distillation the scraps usually give the following results in 100 parts:

| | |
|---|---|
| Zinc | 67·3 |
| Mercury | 4·3 |
| Dross and loss | 28·4 |
| | 100·0 |

The composition of the zinc left after distillation is given as—

| | |
|---|---|
| Zinc | 90· |
| Iron | 2·56 |
| Lead | 6· |
| Copper | 1·44 |
| | 100·00 |

COMPOUND CELL PROCESS.—Another method of economizing power was proposed in what is termed the *compound cell system*, by which it was said that the electricity passing through a series of cells would be able to produce the same quantity of work in every cell with no more cost. This plan may be stated thus:

A is a Smee's battery; the wire Z is conducting the electricity to

* Memoirs and Proceedings of the Chemical Society, vol. ii. page 393.

the compound trough which is composed of a series of water-tight cells, as *a a*, and is connected with a piece of copper C, forming a positive electrode; in the same cell, and facing this electrode, is a medal, connected by a copper wire to a piece of copper placed in the second cell, opposite which is another medal connected in the

Fig. 575.

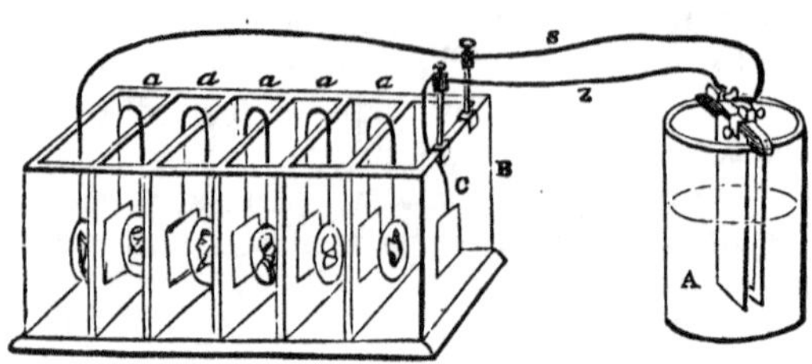


same manner with another piece of copper, and so on through the series, which terminates with a medal attached to the wire of the battery. The electricity from the battery passes through all these cells, and reduces its equivalent in each cell. Thus the reduction of 32 grains of zinc in the battery would deposit 32 grains of copper multiplied by 6 times, or as many times as there are cells.

This is correct in principle, and at first sight seems to be exceedingly economical; but it is not so, for every cell adds so much to the resistance of the current, that intensity batteries must be used; so that, supposing we have a compound cell of six divisions, in which are placed six separate medals, it would require a battery of six pairs of plates to give intensity sufficient to overcome the resistance, and the same number of medals could be made of the same weight by six separate zincs, and in less than half the time they could be made by this arrangement, and with a less destruction of zinc. For large operations, where the articles receiving the deposits and the electrode are necessarily a good way apart, the process is altogether impracticable in a commercial point of view. This is one of the remarkable instances where theoretical possibility and commercial economy are at variance.

Effects of Resistance.—At page 551 we mentioned, that if a single cell deposits 100 grains in a given time, and it be converted into a battery having the two electrodes in a solution of sulphate of copper, there will only be deposited in the same time 30 grains. This is caused by the extra resistance which the solution between the two electrodes, in the decomposition cell, offers to the passage of the electricity, the amount of which corresponds to the amount deposited—the latter depending upon the former.

If we take two small plates of copper and zinc amalgamated, and place them in dilute sulphuric acid, in contact, but not so close as to prevent the gas evolved from the copper plate to escape, and allow them to remain until there have been dissolved from the zinc

100 grains, and we call this the measure of the maximum amount of electricity which that surface of zinc and copper can give out in the time taken to dissolve the 100 grains: then, if the two metals in the acid be separated one inch, being connected by a wire or slip of copper above the liquid, and kept in action the same length of time as the former, there will be dissolved from the zinc only about 56 grains. If the wire in connection with the zinc and copper be extended and cut in the middle, and have a piece of copper attached to each of the same size as the zinc plate in acid, and these be placed in another vessel containing a solution of sulphate of copper (as Fig. 544), and put an inch apart, and the whole kept in action the same length of time as before, it will be found in this case that only 10 grains of zinc are dissolved. From these experiments we see that the resistance of the one inch of acid between the zinc and copper in the battery, and the one inch of solution of sulphate of copper in the second or decomposition cell, is 90 or nine-tenths—only yielding one-tenth of the electricity which the zinc and copper are capable of giving.

INTENSITY.—If we now take another zinc and copper plate of the same size as the former, and arrange them in the acid solution, and connect them with the copper plates in the decomposition cell, as shown in Fig. 544, and keep them in action the same length of time as in the former experiments, there will be dissolved from the zinc about 19 grains, and deposited upon the copper plate attached to the zinc in the decomposition cell 18 grains of copper.

Fig. 576.

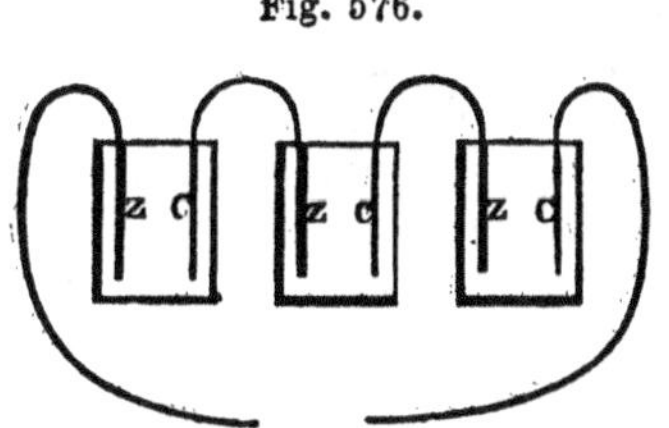


If three zincs and coppers be arranged as described and placed in the acid, there will be dissolved from the zinc plate 26 grains, and deposited upon the copper 25 grains. If six pairs zinc and copper be arranged as above and placed in acid, there will be deposited 36 grains of copper, which we will also take as the measure of what is dissolved from the zinc; and if nine pairs of zinc and copper be used, there will be deposited 43 grains, and so on until the quantity dissolved from each zinc, or deposited on the copper plate be 100, equal to that obtained by the close contact of the zinc and copper in acid, which will require upwards of 30 pairs of zinc and copper. It must be borne in mind that the same quantity of zinc will be dissolved from every plate in the arrangement. Thus, in nine pairs where 43 grains were deposited, there would be dissolved from every zinc in the battery 43 grains.

It will now be apparent that the use of several pairs in the battery is to overcome resistance, by which quantity is gained at the same time up to a given point; but quantity gained by this means is expensive. The 10 grains deposited by the single pair of zinc and copper only required 10 grains of zinc, but the 43 grains by the nine pairs would require 405 grains of zinc to be dissolved.

RELATIVE INTENSITY OF BATTERIES.—Different batteries have different degrees of power to overcome resistance—greater intensity. The following experiments will illustrate this: A single pair of a Wollaston's, Smee's, and Grove's batteries were fitted up as nearly equal in circumstances as the different arrangements would allow—each exposing the same surface of zinc, and connected with electrodes placed in a solution of sulphate of copper, first 1 inch, then 2 inches, 3 inches, and 4 inches apart—half an hour in each. They were then reversed, beginning with the electrodes at 4 inches and coming to 1 inch. These experiments were repeated several times, and a mean of the whole taken. The results were:

| Deposited— | Wollaston. | Smee. | Grove. |
|---|---|---|---|
| Electrodes 1 inch . . . | 8·8 grains | 12·0 | 31·0 |
| 2 inches . . | 6·6 " | 6·8 | 26·0 |
| 3 inches . . | 4·7 " | 6·0 | 17·0 |
| 4 inches . . | 3·0 " | 4·6 | 14·0 |

From this it will be seen that Wollaston's stands lowest in intensity, which is more apparent as the distance of the electrodes is increased. Smee's is one-third more than Wollaston's at 1 inch, and one-half more at 4 inches; while Grove's is three and a-half more than Wollaston's, and two and a-half more than Smee's at 1 inch, but four and a-half more than Wollaston's and three more than Smee's at 4 inches. If we take the mean of these results as a comparison of batteries, their value will stand as under:

One of Grove's equal to three of Smee's,
and to three and three-fourths of Wollaston's.

The following table gives the results of different batteries, arranged in series, kept in action the same length of time, namely, one hour. The battery plates were very small, the electrodes twice the size of the battery plates:

| | One Pair. | Two Pairs. | Four Pairs. | Six Pairs. | Nine Pairs. |
|---|---|---|---|---|---|
| Grove's . . . . . . . . . . | 55 | 72 | 93 | 97 | 98 |
| Daniell's . . . . . . . . . | 15 | 35 | 60 | 77 | 86 |
| Smee's . . . . . . . . . . | 11 | 19 | 29 | 41 | 58 |
| Wollaston's . . . . . . . . | 8 | 15 | 24 | 33 | 48 |

This table gives results approaching to and in principle the same as the others: it will be observed that one pair of Grove's is equal to nine pairs of either Wollaston's or Smee's. It is also worthy of remark, that Grove's increases slowly in quantity above four pairs, the intensity being sufficient at four pairs to overcome the resistance offered to the current of electricity. For ordinary electrotyping, intensity arrangements are unnecessary, except where the article upon which the deposit is being made is of such a character as will not allow the positive electrode to be brought close to it, or when there are deep cut objects, or any circumstance that increases distance and necessitates power to overcome resistance.

Mode of Suspending Objects for Coating.—In beginning to operate in the art of electrotyping, the student often pauses, and asks the question, What is the best position in which a medal should be hung in the solution? Convenience has brought into general practice the suspending of it perpendicularly in the solution, having the positive electrode or pole facing it in a parallel direction; but to this method there are some objections. If, for instance, the porous diaphragm, or single-cell system be used, for obtaining the medals, it is found that upon the lower portion of the medal the deposition is much thicker than upon the upper portion. Indeed, when even ordinary attention is not paid, the lower part becomes not only thicker, but studded over with round nodules of copper, or with lines composed of these nodules, while the upper part remains thin, and is covered over with what is termed the sandy deposit copper, in dark brown grains, capable of being rubbed off with the slightest friction. No doubt this is in a great measure prevented by agitating the solution; but it is inconvenient, and requires constant attention.

Fig. 577.

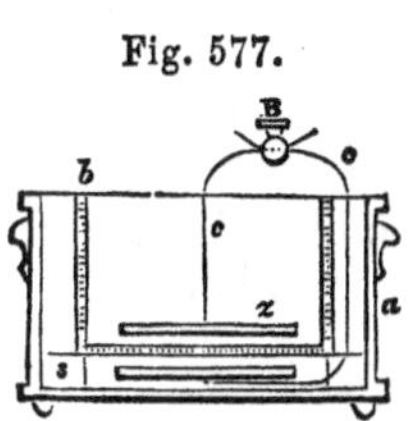

If a separate battery is used, and the deposition of the medal is effected in a separate vessel, by having a copper positive electrode, the same inconvenience takes place to a greater or less extent, according to the distance at which the two poles are placed. These inconveniences are known to all electrotypists, and the cause is ascribed to the different densities of the solution. The reason why the solution becomes of different densities is easily understood in the single cell process, there being no copper pole to maintain the strength of the solution; as it becomes exhausted of copper by the deposition, the lighter portion floats on the top, and the heavier portion remains below; and although crystals of sulphate of copper be suspended in the solution, as they dissolve they sink by their gravity, and cause a flow upon the lower portion of the medal, and consequently a much more powerful deposit. But why the same should take place with a separate battery, where there is a positive electrode of copper being dissolved, just in proportion to the copper extracted from the solution by the medals, was for a long time not known.

NON-TRANSFER OF ELEMENTS.—In a paper read upon this subject before the Royal Society by the late Professor Daniell and Professor Miller, they gave, as the results of their investigations, that certain metals are transferred by the electric current in small proportion, differing in different metals.* About the same time the author had observed, when operating upon the large scale, results which led him to the conclusion, that no metal is transferred in any quantity by the electric current, nor any element taking the position of the metal in an electrotype, but that the acid element was always transferred equivalent to the electricity passing. It was thus shown that during the deposition of metal, say copper, in electrotyping, the acid, when exhausted of the copper at the surface of the medal, is transferred to the positive pole, and dissolves a portion of copper; but this portion is not transferred by the electric current to the medal: hence it will be observed, that the solution next the medal will become exhausted of copper, and will consequently rise to the surface from its greater lightness. There is no doubt a flow of stronger solution in a horizontal direction from the positive pole to the medal, caused by the lighter portion ascending; but this does not mend the evil: the light portion is increasing on the surface, and the whole solution soon becomes of different densities from the surface to the bottom of the medal; and this constant current of the solution flowing up the surface upon which the electrotypist is depositing, causes the lines that are observed in deposits under certain circumstances, and which are sometimes very annoying. If a small hollow be in the mould, or even if a small portion of a plain surface resist, the metal will accumulate round the edge of the resisting portion, giving the deposit an appearance as if made in a flowing stream, like a stone standing up in a current of water. The black point in the centre represents the resisting spot around which the deposit will thicken, causing a ridge of metal to radiate to a point immediately above the resisting portion. These disappointments are much more annoying in solutions of gold and silver than in sulphate of copper, as will be noticed when we come to treat of plating and gilding. A point of grease or dirt, or small hole not cleaned out, hardly visible to the naked eye, will give a very prominent effect upon the plain polished surface of a piece of metal.

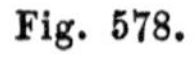
Fig. 578.

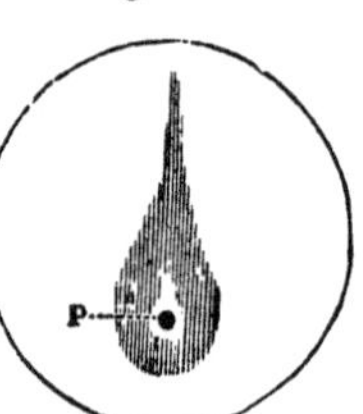


From these observations, the reader will now be able to answer the question—what is the best position to place a medal in the solution? To make it still more apparent, take a glass jar, filled with a solution of sulphate of copper; place a piece of copper upon the bottom of the jar, and suspend the medal at the top,

* Philosophical Transactions, Part 1, for 1844; and Memoirs and Proceedings of the Chemical Society, Vol. iii. page 53.

having their two faces parallel; connect them with a battery; in a short time the solution round the medal becomes exhausted, and even colorless, the medal covered with a dirty brown powder, and no further deposit will take place. But reverse the case; place the medal at the bottom, and the copper positive electrode at the top; the deposition goes on constant and smooth; the solution is maintained in the same condition as it was at the first, there being a constant transfer; the acid is transferred by the current from the medal to the copper pole: the sulphate of copper formed descends by its gravity to the medal. There are, no doubt, a few slight objections to placing the medal under the positive electrode—such as the impurities in the copper getting disintegrated, and falling upon the surface, but a piece of cloth wrapped round the pole prevents this. However, when a fine surface is wanted, care ought always to be taken to have clean solutions filtered, and kept covered from dust; and when the single cell is used, the crystals of sulphate of copper should be suspended in a fine linen bag, or the shelf holding them be lined with linen.

EFFECTS OF DIFFERENCE IN THE DENSITY OF SOLUTION.—Although in principle this is the best method, we believe that very few practice it, because of the trouble attending the arrangement oi the electrodes in this position. When the medals are small the annoyances from unequal density are not so material, but if the surface of the article which is being deposited be large—say eight inches or upwards—the difference in the thickness of the lower and upper portion of the medal is very great. When suspended perpendicularly, they should be shifted several times, making the upper portion the lower, besides occasionally stirring the solution, or shaking the article. Indeed when convenient, the article receiving the deposit should be kept as much in motion as possible, as it regulates the deposit, making it smoother and less brittle.

CRYSTALS OF COPPER ON ELECTRODES.—It will be found, when working with a battery, that the sulphate of copper solution will become stronger round the positive electrode, which is gradually dissolved by the transferred acid. A frequent effect is, that the electrode often gets coated over with crystals of sulphate of copper, which adhere with great tenacity, and stop the electric action. Under such circumstances, it is only necessary to clean the electrode from the crystals and to add a little water to the solution, which will prevent a recurrence of the crystals for a time. But the stirring of the solution occasionally will do much to prevent this crystallization.

# CHAPTER XXVII.

## MISCELLANEOUS APPLICATIONS OF THE PROCESS OF COATING WITH COPPER.

Besides the applications and processes which we have described under the general term of electrotyping, there are various applications of the process of depositing metals upon other substances, which have been, and may be still more usefully applied. We may, at a trifling cost, impart a coating of copper to cornices for decorating buildings, to terra cotta, engravings on wood, etc., etc. Cloth may also be easily covered, and made to assume the appearance of a sheet of copper, having the lightness and pliability of cloth. Lace has been covered with copper, and used for battery plates, and has also been gilt and made into beautiful ornaments. Table-covers with metallic ornaments richly gilt, and book-covers, have all been tried with more or less success, although they have not yet been profitably produced.

Coppered Cloth.—Ordinary cloth, covered with copper, was prepared a few years ago in considerable quantities for the covering of roofs, wagons, etc.; but the necessary price precluded its use when competing with the ordinary materials for these purposes, although it possesses many eminent qualities for some of these uses—such as forming fire-proof covers to shelter wagons from the sparks discharged by a locomotive. The choice of the kind of cloth was another difficulty; linen was too expensive, and required a good coating of copper to make it water-tight; the best substance was a felted cotton with India-rubber, but after a few months exposure the India-rubber in the cloth decomposed. The operations of coating cloth with copper were the same as described for the wax medals: the cloth was brushed over with a polish of black lead, and then stretched upon a frame of wood having a copper band round it, in which were placed small hooks or pins, and the cloth attached to these. A vat, four feet deep and twelve yards long, was made of brick and cement; this was divided lengthwise by a wooden frame with panes the same as a window, which were filled in with unglazed earthenware plates, cemented by marine glue, and the whole made water-tight. Into one division of the vat were placed the dilute acid and sheets of zinc; in the other the solution of copper, in which was placed the cloth upon the frame. The arrangement was so perfect that we have often seen pieces of cloth, twelve yards long by one yard wide. completely covered with copper in one hour. The result of many trials was, that one pound weight of copper gave a perfect solid covering to twenty superficial square feet of cloth.

A similar thickness is quite sufficient for other surfaces for mere exposure to the atmosphere, such as wood-work, cornices, etc., and may be produced at the same rate, about 2s. 6d. per pound of copper.

Besides these applications, many others have been suggested and tried with variable success. Some have probably been abandoned too soon, others have had both capital and talent applied, and success is yet to come. We shall only name a few of these applications.

CALICO PRINTERS' ROLLERS.—So early as 1841 active means were tried to apply the electro-deposition of copper to the preparation of rollers for printing calicoes, both by depositing the copper upon wax or other moulds, to make an entire roller of copper, or to deposit a surface of copper on other metals, such as iron or brass; but none of them have yet succeeded. To make an entire roller is much more expensive, without an equivalent advantage over the ordinary method of casting, rolling and boring. To deposit a layer of copper on iron is attended with many practical difficulties, both in protecting the iron from the acid solution for so long a time as is required to deposit the proper thickness, and in securing the adhesion of the two metals during the subsequent operations. It requires a deposit of about a quarter of an inch in thickness to allow for turning before engraving. There is then the annealing to soften the copper, etc., which interferes with the adhesion of the two metals, probably from their different rates of expansion, and other causes. Similar objections may be made to the coating of brass rollers with copper. Numerous and varied have been the experiments made, but all without success; nevertheless we have every confidence that means will be obtained for producing rollers by the electro process.

ETCHING OF ROLLERS.—Another application of the process to printers' rollers was to plate the surface of the roller with silver for the purpose of etching. The engraving is then made through the silver coating; the roller is next passed through nitric acid, which acts upon the exposed copper, the silver taking the place of varnish in ordinary etching: but practical difficulties have caused the abandonment of this application also.

PRINTING.—The electro-metallurgical process has been applied to many operations in ordinary printing. Mr. Warren de la Rue has been eminently successful in many of these applications. It has also been applied to plates for printing music, and for embossing soft materials, such as leather. By depositing a sheet of copper upon a skin of morocco leather, it may be used for imparting an impression to other skins of leather, giving them the appearance of fine morocco.

The printing of music has also been successfully done by electrotyping the plate from a stereotype cast. The same may be done from ordinary stereotype plates.

GLYPOGRAPHY.—A process which Mr. Palmer, the inventor, named Glyphography, has been one of the most successful attempts to apply the electrotype to the art of engraving. The principle of the invention consists in depositing copper in the grooves or engravings made in a layer of some soft substance spread on a sheet of copper, and covering the whole with a sheet of electrotype copper. The counterpart of the engraving thus produced is used for printing from in the same manner as letter-press printers' types or woodcuts. It may therefore be called a mode of stereotyping, with this difference, that it is made directly from the drawing by the artist. The drawing, however, must be made in a particular way, which, with the other necessary manipulations, is thus given by Mr. Palmer:*

"A piece of ordinary copper plate, such as is used for engraving, is stained *black* on one side, over which is spread a very thin layer of *white* opaque composition, resembling white wax both in its nature and appearance. This done, the plate is ready for use.

"In order to draw properly on these plates various sorts of points are used (according to the directions here given), which remove, wherever they are passed, a portion of the white composition, whereby the blackened surface of the plate is exposed, forming a striking contrast with the surrounding white ground, so that the artist sees his effect at once.

"The drawing, being thus completed, is put into the hands of one who inspects it very carefully and minutely, to see that no part of the work has been damaged or filled-in with dirt or dust; from thence it passes into a third person's hands, by whom it is brought in contact with a substance having a chemical attraction or affinity for the remaining portions of the composition thereon, whereby they are heighted *ad libitum*. Thus, by a careful manipulation, the *lights* of the drawing become thickened all over the plate equally, and the main difficulty is at once overcome. A little more however remains to be done. The depth of these non-printing parts of the block must be in some degree proportionate to their width, consequently the larger breadths of *lights* require to be thickened on the plate to a much greater extent in order to produce this depth. This part of the process is purely mechanical and easily accomplished.

"It is indispensably necessary that the printing surfaces of a block prepared for the press should project in such relief from the block itself as shall prevent the probability of the inking-roller touching the interstices of the same whilst passing over them. This is accomplished in wood engraving by cutting out these intervening parts, which form the lights of the print, to a sufficient depth; but in glyphography the depth of these parts is formed by the remaining portions of the white composition on the plate, analogous to the thickness or height of which must be the depth on the block, seeing that the latter is, in fact (to simplify the matter), a *cast or reverse* of the former. But if this composition were

* Glyphography, or engraved drawing, for printing at the type press after the manner of woodcuts, 1844.

spread on the plate as thickly as required for this purpose, it would be impossible for the artist to put either close, fine, or free work thereon; consequently the thinnest possible coating is put on the plate previously to the drawing being made, and the required thickness obtained ultimately as described.

"The plate thus prepared is again carefully inspected through a powerful lens, and closely scrutinized to see that it is ready for the next stage of the process, which is to place it in a trough and submit it to the action of a galvanic battery, by means of which copper is deposited into the indentations thereof, and, continuing to fill them up, it gradually spreads itself all over the surface of the composition until a sufficiently thick plate of copper is obtained, which on being separated will be found to be a perfect cast of the drawing which formed the *clichée.*

"Lastly, the metallic plate thus produced is soldered to another piece of metal to strengthen it, and then mounted on a piece of wood to bring it to the height of the printer's type. This completes the process, and the glyphographic block is now ready for the press.

"It should however have been stated previously, that if any parts of the block require to be *lowered,* it is done with the greatest facility in the process of mounting."

This process has however not come into much use as a substitute for wood engraving, in consequence of the impossibility of finding a suitable varnish for the use of the artist or engraver. It has, in fact, given way to another process, also embraced in Mr. Palmer's patent, which is worked thus: A copper plate is etched by the process commonly employed by engravers, the lines being cut into the copper with a bold stroke. The lines are then bitten deeper by nitric acid. The etching is made *direct,* not *reversed,* as it is upon a plate that is to be worked at the copperplate press. When the engraving is ready, the etching varnish through which the drawing is cut is covered with a conducting substance, and an electrotype plate is deposited upon the etching. When this is removed from the mould it requires to be trimmed, for it is impossible to etch a plate, or to bite the etching, so that all the lines shall be exactly of the same depth. To remedy this the face of the electrotype is levelled by grinding and burnishing. The following instructions for artists are published by the patentee:

INSTRUCTIONS ON GLYPHOGRAPHY FOR THE AMATEUR.—"The amateur must remember that he is producing a work of art for the surface press, and not for copperplate printing.

"The drawing or etching should not be made with lines of equal thickness in all the tints. If it is so treated with a *thick* line, and if the cross hatching be kept of the same strength as the principal line, it will appear like a coarse pen-and-ink drawing. If it is treated in the above manner with a *fine* line, and the work laid very close, it will have the appearance of one of the old etchings. The amateur, therefore, will do well to remark, that it is only by a judicious mixture of bold and delicate work that beauty of style can be obtained; and as the darkest shades are generally foremost, and become gradually lighter to the distance, so that the darkest or nearest tones should generally be formed by the boldest work, and gradually increase in delicacy to the offscape.

"Etching is a process nearly resembling drawing with a very fine pen or pencil, and should be proceeded with as follows:

"Having obtained a polished copper-plate with an etching ground properly laid, proceed to put your design upon the plate.

"If it is a print or miniature that is being copied, you must make a sketch or tracing of the same with a black lead-pencil: it must then be traced on to the plate, remembering always, that the proof from the block will be in the same position as the etching; and that nothing must be etched or written backwards, as for the ordinary copperplate printing.

"In order to trace the object on to the plate, take a piece of transfer paper,* place it face downwards upon the plate, secure the corners with a piece of wall wax or paste, or hold it steadily down, if there is not much to trace; then place on your sketch or tracing, go over the outline with your etching-needle or a very hard black lead-pencil, removing a corner at a time to see that all is correctly transferred, and nothing omitted, or that the outline be not too heavy and thick, in which case you must trace lighter.

"Having thus got your subject, as it were, sketched upon the plate, proceed in all respects with your etching-needle as if making a drawing with a black lead-pencil, only working more firmly, taking care always slightly to cut the copper.

"Be careful not to try to form the dark touches and the *black* parts of the subject with a number of lines crossing and recrossing each other, but scrape them away entirely with the point of your penknife, or any other convenient instrument.

"In commencing the etching of a view, it is usual to begin with the offscape, etching the same as neatly and as close *as the nature of the printing will admit*, working more firmly and boldly in every progressive tone, until you reach the foreground. In portraits it is usual to commence with the eye; and in draperies at the top, working downwards.

"Owing to the great difference between surface and copperplate printing, depth of tone should be sought as much from the breadth or thickness of the lines, as from laying them close together; and on the contrary, lightness of tint must be obtained by the distance of the lines from each other, as well as from their delicacy.

"If you make a false line, or wish to efface any portion of the work, a little Brunswick black (which can be procured at most oil and color shops), spread thinly, may be used to stop it out; or rub

---

* To prepare the transfer paper, take some thin post or tissue paper, rub the surface well with black lead, vermilion, red chalk, or any coloring matter: wipe this preparation well off with a piece of clean rag, and it will be ready for use.

a little of the superfluous ground from the side of the plate with a camel's hair pencil and turpentine: when this is dry the work can be re-etched and finished at pleasure."

This last process has afforded some excellent work in the shape of maps, among which we may cite the *Penny Atlas*, published by Messrs. Chapman and Hall. Among subjects of a more picturesque nature executed by glyphography, we may instance Mr. George Cruikshank's etchings of *The Bottle*. These are sufficient to show that the art of electrotyping engravings, though yet in its infancy, promises to be hereafter of importance in the fine arts.

COPYING OF COPPERPLATE ENGRAVINGS.—Copperplate engravings, of all sizes, and of every degree of excellence, have been copied by electrotype. The process is exactly the same as that of making a copy of a penny-piece, as described at page 533; namely, an electrotype mould is first made in copper, on which, of course, the engraving appears in relief; upon which mould any number of electrotype copies of the copperplate engraving may be deposited successively. The duplicates thus made are accurate copies of the original engraving; but they are rapidly worn away by the friction they undergo in the ordinary process of copperplate printing. The process has therefore not displaced the use of engravings on steel.

COATING OF GLASS AND PORCELAIN.—This is done by putting a fine coating of copal varnish over the glass, then black-leading it, and depositing the copper. Another method has been proposed, namely, to make a varnish of two parts asphaltum and one part mastic, by fusing these together, and, when cool, dissolving the mixture in spirits of turpentine to a syrup consistence. To prevent the deposit coming off the glass, the vessel is first corroded by the fumes of hydrofluoric acid. A solution of gutta percha or benzole has also been proposed as a varnish for fixing on the black-lead and deposit.*

Retorts, basins, and other chemical vessels, are sometimes covered with copper for their protection during boiling and evaporation. China saucepans have also been made and covered with copper to take the place of tinned copper vessels, but the adhesion of the metal upon these substances, even when we attempt to secure it by the means above referred to, is never so perfect but that after a short use the deposit of copper loosens from the vessels. There is then great liability for liquids to get between the coating and the vessel, and when heat is afterwards applied these liquids saturated with verdigris boil out. Consequently such coverings are not well adapted either for culinary purposes or delicate chemical operations. They have, notwithstanding, been highly recommended, and the practice of covering the bulbs of large plain retorts, etc., may be useful in a few large manufacturing operations, but our experience is certainly not favorable to their general use.

Mr. John Ridgway, of Cauldon-place, Staffordshire, china manu-

* Progress of General Science, vol. ii.; and Pharm. Journal, vol. viii.

facturer, has recently patented certain improvements in the method or process of ornamenting or decorating articles of glass, china, earthenware, or other ceramic manufactures. In the specification of his patent, just enrolled, Mr. Ridgway states that his first object is to apply a new glaze, which shall enable the metallic coating to adhere firmly, by capillary attraction, and give affinity for copper as a first coating. In pursuance of this, he first submits the article to an alcoholic solution, or a gelatinous solution. He then brushes over it an impalpable powder, composed of half carburet of iron, and half sulphate of copper. The article thus treated is then to be corroded by the fumes of hydrofluoric acid. The article is then to be smoothed, by brushing it over with silver sand, or by the scratch-brush; but when the shape and nature of the article will not admit of this, it is to be plunged into a liquor, consisting of 6 quarts sulphuric acid, 4 quarts aquafortis, ¾ oz. muriatic acid, and 6 quarts water. Grease is to be carefully removed from the article, and a thin film of mercury is to be applied. The solution of copper consists of 1 sulphate of copper, and 4 filtered water. Suitable solutions for silvering or gilding are to be applied, in accordance with the practice of electrotyping. The claim is not to the solutions the coating as such, but to the application of "electrotyping," or electro-metallurgy, to the objects stated in the title, provided the articles be so prepared as to allow them to combine from an alloy with them.

On Galvanic Soldering.—Among the many applications of the deposition of metals, there is one we have been often asked about, namely, if it would not be possible to solder different metals together by that process. The following article, which is taken from the Technologist, will give a full reply to all who may be still inquiring for this application:

"Under the name of galvanic soldering, a process is known by means of which two pieces of metal may be united by means of another metal, which is precipitated thereon through the agency of a galvanic current. This mode of soldering by the 'wet method' has been often recommended in various periodicals relating to the industrial arts; but it has been objected that, practically speaking, the union between two pieces of metal could not be effected by means of a metal precipitated by galvanic agency. In order, however, to arrive at a definite conclusion upon this question, M. Elsner undertook the following experiments, the results of which are in favor of the practical use of the operation of soldering by galvanic agency. In conducting these experiments, the kind of battery known as Daniell's 'constant battery' was employed; and upon the end of the copper wire, which formed the negative electrode, a strong ring of sheet-copper was placed. This ring was cut asunder at one point, and the distance left between the several parts was about the sixtieth of an inch. At the end of a few days (during which time the exciting liquors were several times renewed) the space in the severed portion of the ring was completely filled up with copper regulus, which had been precipitated; and on partially cutting with a file through the part thus filled up, and examining it with a lens, it was observed to be very equally filled with solid and coherent copper.

"Another copper ring was then cut into two parts, and the two semi-angular segments thus obtained were placed with the faces of the sections opposite each other, and submitted to the action of a galvanic current. At the end of a few days, the segments were united by the copper precipitated, thus forming again a complete ring. It was also found in this case, on removing with a file a portion of the thickness of the ring at the points of contact, that the spaces had been completely filled up by copper galvanically precipitated, which had united the whole. On observing these points carefully with a lens, the regular deposition of the copper could be readily traced between the formerly separated portions of the ring.

"A third experiment was made in the following manner: Two strong rings of sheet-copper were laid with their freshly-cut faces one upon another, so that the two rings constituted a cylinder. These rings were surrounded by a band of sheet-tin which was coated with a solution of wax, so that the two rings were equally surrounded by a conducting material. Thus disposed, these rings were attached to the negative wire of the battery and immersed in the bath of sulphate of copper. At the end of a few days the interior surface of the rings was covered with precipitated copper, and between the contact surfaces of the two rings copper was also precipitated. These rings had only been submitted to the galvanic current to such an extent as to cover their interior surface with a thin coating of precipitated copper, and yet they were already completely re-united, and formed a cylinder consisting of a single piece. The exterior conducting covering, consisting of a sheet of tin, was of course removed before testing the cohesion or persistence of the galvanic precipitate. It may be remarked that these rings, after being for a certain time in contact (during the galvanic action), together with the plate of copper upon which they rested, became so encrusted with precipitated metallic copper that some force was found necessary to effect their detachment from the copper wire.

"There would appear to be no doubt, then, according to the results obtained in the preceding experiments, that two pieces of metal may be firmly united by means of galvanically-precipitated copper: in a word, that soldering by galvanic agency is perfectly practicable. It will therefore be possible to firmly unite the different parts of a large piece of metal, and to make a perfect figure of them by galvanic precipitation of a metal (copper, in ordinary cases). If solutions of salts of gold or silver were employed in as concentrated a form as those of copper above mentioned, there is reason to believe that galvanic soldering would also result. In fact, M. de Hackewitz states that in some experiments on a larger scale which he undertook, to obtain hollow figures by galvano-plastic means, he had remarked that galvanic union often took place between the pieces operated upon. M. Elsner states that while conducting the experiments above-mentioned, he remarked that by employing too powerful a current, the negative electrodes of copper, and even the plate of copper, and ring of the same metal resting thereon, became covered with a deep brown substance, in the same manner as this occurs under similar circumstances in galvanic gilding, as is well known. After several unsuccessful attempts to prevent the formation of this brown coating, M. Elsner found that it was possible to remove it entirely on immersing the articles covered therewith during a few seconds in a mixture of sulphuric and nitric acids. By this means the precipitated copper was made to assume its natural red color. The possibility of practically effecting the operation of soldering by galvanic agency may be explained in a few words, in a theoretical point of view. The article is in fact in an electro-negative state of excitation, whilst the zinc operates positively. The result is, that the faces which are placed opposite each other, when the ring has been cut, are negative; that is to say, in an electric condition of the same denomination. During the progress of the electrolytic decomposition of the metallic salt in solution (sulphate of copper in the above case), the electro-positive molecules of copper which are detached simultaneously arrange themselves upon the two opposite faces, and in the direction of the break. Now, from the moment that these molecules are deposited, they constitute with the piece a homogeneous mass, and from that time act negatively upon the copper which is contained in the solution, and again precipitate copper in the form of regulus. This method of operation continues until the space which existed between the two separate pieces of metal is filled up with metallic copper—in fact the layers of copper which become deposited in an equal manner upon the contiguous faces of the metal gradually diminish the distance which separated the latter, until at length the metallic layers which cross in the opposite direction meet each other; the result being, that the whole of the break which originally existed between the faces will have disappeared and become filled up with copper.

"With respect to the solidity (the degree of cohesion) of the galvanic soldering, it is the same as that of copper or other metal precipitated by galvanic agency. It will moreover be well understood that too energetic galvanic excitation must have an injurious influence upon the cohesion of the metal precipitated; and in this case precisely the same phenomena will be observed as those which have long manifested themselves in ordinary galvano-plastic operations."—L. ELSNER, *Technologist.*

We mention another application of the electro deposition, which might be extended.

GALVANO-PLASTIC NIELLO.—Niello, a peculiar style of enamelling, consists in engraving or stamping figures on a plate of silver or gold, and then filling the incised lines or impressed pattern with a sort of enamel—differing, however, from true enamel—which is a kind of glass, by being formed of a mixture of the sulphurets of lead, silver, and copper. This mixture is of a black color—hence the name niello, from nigellum, derived from niger, black—and when melted into the intaglio parts of a plate gives it somewhat the appearance of an inked engraved copperplate. A new kind of niello work has lately been introduced on the Continent, in which, however, the figures are not produced by an enamel of sulphuret of silver, as in the true niello, but by a different-colored metal; thus on a plate of gold may be produced fine engravings, the lines of which are in silver, and so on. This can be effected in two ways: first, by covering the plate to be ornamented with a varnish exactly as is used in etching; the pattern or ornament is then to be engraved on this varnish, and the metallic surface etched out to the proper depth. The engraved plate is to be placed in a solution of the metal intended to form the pattern, and a deposit allowed to form in the usual way adopted in all galvano-plastic works. When the intaglio lines have been completely filled up by the deposited metal, the plate is removed from the solution and ground, when the pattern will be fully developed. The second method consists in sketching the ornament on a sheet of paper with lithographic ink, placing this, with the side upon which the drawing was made, upon a plate of silver or other metal to be ornamented, and pressing them together; the paper is now removed with water, slightly acidified, leaving the ink adhering to the plate, which is to be sprinkled with sand. When the ink has fully dried, the sand is blown away; the plate is placed in a solution of the metal which it is intended should form the ground, and put in connection with the battery. By this means a deposit will be formed over the whole surface, except the parts protected by the ink. On the removal of the latter with alcohol or spirits of turpentine, etc., the original metal will be exposed, forming a pattern. Many highly ornamental and useful applications might be made of these processes, especially in the manufacture of church furniture. Instead of simply engraving the name and legend upon pieces of plate presented to persons, it might be put in in letters of gold at very little more expense.

---

## CHAPTER XXVIII.

### BRONZING.

WE have already mentioned that when a medal has been made from a metallic mould, protected by a little wax dissolved in turpentine, it retains its bright copper lustre for a long time, even when exposed to the air; but generally the copper medals and other objects are very liable to tarnish, for which reason it is usual to give them a coating of bronze, that they may acquire a permanently agreeable appearance.

BROWN BRONZES.—Bronzing is effected by several very simple methods, the most common of which is the following:

Take a wine-glass of water, and add to it four or five drops of nitric acid; with this solution wet the medal (which ought to have been previously well cleaned from oil or grease) and then allow it to dry; when dry impart to it a gradual and equable heat, by which the surface will be darkened in proportion to the heat applied.

ANOTHER METHOD.—Make a thin paste of crocus and water: lay this paste on the face of the medal, which must then be put into an oven, or laid on an iron plate over a slow fire; when the paste is perfectly reduced to powder, brush it off and lay on another coating; at the same time quicken the fire, taking care that the additional heat is uniform; as soon as the second application of paste is thoroughly dried, brush it off. The medal being now effectually secured from grease, which often occasions failures in bronzing, coat it a third time, but add to the strength of the fire, and sustain the heat for a considerable time: a little experience will soon enable the amateur to decide when the medal may be withdrawn; the third coating being removed, the surface will present a beautiful brown bronze. If the bronze is deemed too light the process can be repeated.

Another very simple method is this: after the medal is well cleaned from wax or grease, by washing it in a little caustic alkali, brush some black lead over the face of it, and then heat it in the same way as described for crocus; or a thin paste of black lead may be used, and the processes already referred to be repeated until the desired brown tint is obtained. In this kind of bronze a little Hematitic iron ore, which has an unctuous feel, may be brushed over the face of the bronze, by which a beautiful lustre is imparted to it, and a considerable variety in the shade may be obtained. In the brown bronzes the copper is slightly oxidized on the surface.

BLACK BRONZES.—A very dark colored bronze may be obtained by using a little sulphuretted alkali (sulphuret of ammonia is best). The face of the medal is washed over with the solution, which should be dilute, and the medal is to be dried at a gentle heat. It should afterwards be polished by a hard hair brush. Sulphuretted hydrogen gas is sometimes employed to give this black bronze, but the effect of it is not so good, and the gas is very deleterious when breathed. In these bronzes the surface of the copper is converted into a sulphuret.

Many metallic solutions, such as weak acid solutions of platinum, gold, palladium, antimony, etc., will impart a dark color to the surface of medals when they are dipped into them. The medal after being dipped into the metallic solution is to be well washed and brushed. In such bronzes the metals contained in the solution are precipitated upon the face of the copper medal, which effect is accompanied by a partial solution of the copper.

GREEN BRONZES.—Green bronzes require a little more time than those already described. They depend upon the formation of an acetate, carbonate, or other green salt of copper upon the surface of the medal. Steeping for some days in a strong solution of common salt will give a partial bronzing which is very beautiful, and if washed in water, and allowed to dry slowly, is very permanent. Sal ammoniac may be substituted for common salt. Even a strong solution of sugar, alone, or with a little acetic or oxalic acid, will produce a green bronze; so also will exposure to the fumes of dilute acetic acid, to weak fumes of hydrochloric acid, and to several other vapors. A dilute solution of ammonia allowed to dry upon the copper surface will leave a green tint, but not very permanent.

Electrotypes may also be bronzed green, having the appearance of ancient bronze, by a very simple process: take a small portion of bleaching powder, place it in the bottom of a dry vessel, and suspend the medal over it, and cover the vessel: in a short time the medal will take on a green coating, the depth of which may be regulated by the quantity of bleaching powder used, or the time that the medal is suspended in its fumes: of course any sort of vessel, or any means by which the electrotype may be exposed to the fumes of the powder, will answer the purpose: a few grains of the powder is all that is required. According as the medal is clean or tarnished, dry or wet, when suspended, different dints with different degrees of adhesion will be obtained.

The green bronzes are generally applied to figures and busts.

These directions and hints will enable the student to vary his bronzes. Practice will give him perfection, and enable him to fix upon that which best pleases his taste. Scarcely two electrotypists agree upon the same method of bronzing, but differ in some little details of practice, or on some point of taste. Each prefers the plan that has given to him his best results, and which he can hardly impart by description to another.

Should the electrotypist wish to coat the copper medal with another metal, as silver and gold, directions will be given under plating and gilding how to proceed to effect his purpose.

## CHAPTER XXIX.

### DEPOSITION OF METALS UPON ONE ANOTHER.

COATING OF IRON WITH COPPER.—Besides making articles of solid copper, we may at a small cost give a coating of copper to another metal, such as iron, which if kept in a dry place, will retain the appearance of copper for any length of time. But in covering iron with copper, or any one metal with another, great care must be taken that a proper kind of solution be used.

It is a familiar fact, that if a piece of iron, such as the blade of a knife, be dipped into a solution of sulphate of copper, it receives a coating of that metal. This is often described as the result of galvanic action, but there is no more galvanic action in this than in any ordinary chemical combination; it is simply a case of chemical substitution; the acid that is in union with the copper having a stronger attraction for iron, leaves the copper, and combines with the iron: the copper is left on the surface of the iron, but the two metals not having sufficient polar attraction to cause them to adhere so firmly as to exclude the action of the acid, the copper is undermined, and falls to the bottom of the solution as a powder. After some copper has fallen upon the surface of the iron, local galvanic action is induced between it and the iron; but this secondary action is altogether distinct from that which first takes place.

Any solution that has the power to give a metallic coating to a metal when dipped into it, should not be used to coat that metal by electricity.

The attraction of the common mineral acids for the ordinary metals is as follows: Zinc, iron, copper, nickel, silver, gold, platinum.

If the metal to be deposited be copper held in solution by an acid, say sulphuric acid, then iron or zinc cannot be coated with copper from this solution; the acid having a greater attraction for these metals, will leave the copper and combine with them as described above: but if the metal to be coated be any of those under copper, in the above table, then no chemical action will take place, and no deposit will be made, except as the effect of the electric current introduced by the battery. This we believe is the cause why De la Rive, Spencer, and others, failed, at an early stage of the art, in their experiments in plating and gilding, as they employed acid solutions, which are quite impracticable when used for depositing upon inferior metals. Under these circumstances, other solvents for the metals must be used, which have a different relative attraction for the metals than the acids have. The substance first applied for this purpose is, after sixteen years experience, still found to be the best—namely, cyanide of potassium.

Cyanide of Potassium.—This substance may be prepared by exposing ferrocyanide of potassium (yellow prussiate of potash) to a red heat in an iron crucible; then pounding the mass, and boiling it in alcohol of about spec. grav. 0·900: cyanide of potassium crystallizes on cooling the resulting solution. It is now, however, almost universally prepared for electro-metallurgical purposes, by a process which was first suggested by Messrs. F. and E. Rodgers, but afterwards more fully explained by Professor Liebig, and hence called "Liebig's Process:" it is at once both simple and easy of performance.

Ferrocyanide of potassium, pounded fine, is dried over a slow fire (we have found an iron plate, or clean shovel, to serve the purpose very well): it must be constantly stirred to prevent its forming a cake upon the hot iron; when perfectly free from moisture, 8 parts must be thoroughly well mixed with three parts of carbonate of potash, also well dried: put a cast-iron crucible into the fire, and, when it is red hot, nearly fill it with the mixture, and keep up the heat by occasional augmentations of fuel: the crucible should be kept covered as much as possible. In a short time the whole fuses into a beautiful liquid with the evolution of gas. It should be kept in this state for 10 or 15 minutes, being occasionally stirred with an iron rod: the portion adhering to the red should be examined from time to time, and when the liquid on it cools white, it is an indication that it is ready to be removed from the fire; but the first time a cast-iron crucible is used, this test will not be so accurate, the salt having then a light gray color. When the crucible is removed from the fire, it should be placed upon a stone, the mass stirred, and then allowed to settle for a short time, after which the clear, or liquid part, is to be poured off into a clean iron vessel. The sediment should be scraped clean out of the crucible while it is hot, as the crucible will do to use again several times; but if the mass at bottom be allowed to cool it will be difficult to remove it from the crucible afterwards. The clear liquid poured off is cyanide of potassium, having from 25 to 30 per cent. of cyanate of potash, and other impurities generally contained in commercial yellow prussiate of potash: 80 per cent. of cyanide of potassium is the greatest proportion that this process can give. We have occasionally obtained it at 78 per cent. from commercial materials, but more generally at 70 and 72 per cent; and we have found cyanide of potassium in the market containing as little as 49 per cent. of pure cyanide.

The results of the manufacture of this salt on a large scale, from the ordinary materials of commerce, show that 55 lbs. of yellow prussiate, dried as directed above, yield 48 lbs.; and 19 lbs. of carbonate of potash give 18 lbs. of dry salts; in all 66 lbs. of the proper mixture. The crucible used was of this shape, Fig. 579, capable of holding from two to three pints; in general two of them were used up in making the above quantity of cyanide, even when great care was taken. One great cause of the crucible giving way is the depth of the fire, and openness of the bars of the grate. The bottom of the crucible, between each pair of bars, fuses from the great heat concentrated near the opening. To remedy this evil, a square tile of fire-clay should be laid upon the bars upon which the crucible is to rest. The tile must not cover all the bars, else the draught will be stopped—an equal space must be left at each side of the tile, which will preserve a regular heat around the crucible.

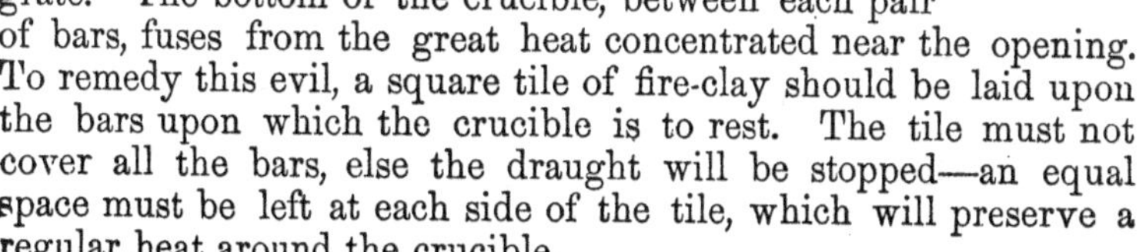
Fig. 579.

The quantity of clean cyanide of potassium obtained from the above quantity of materials was about 38 lbs.; the sediment scraped out of the crucible, being put into water, yielded about 6 lbs. more in solution, but of inferior quality—good enough, however, for precipitations, the cleaning of silver, and other general purposes in the factory.

It may be mentioned that in these operations the crucible is never allowed to cool, but as soon as the sediment is scraped out, it is again put into the furnace. If the iron sediment is not well cleaned out, it imbibes oxygen rapidly, and the charge next taken from the crucible will have an excess of cyanate of potash, besides lessening the capacity of the crucible. Generally speaking, however, even when the utmost care is taken, the last charge has more cyanate of potash than the first.

Cyanide of Copper.—To prepare copper solutions by means of cyanide of potassium, for covering iron and other positive metals, there are several methods.

*First Method.*—To a solution of sulphate of copper, add by degrees a solution of cyanide of potassium, which will give a yellowish green precipitate, with slight effervescence. There will be evolved a gas, having a most pungent odor, to prevent the inhalation of which the most watchful carefulness has to be exercised, as it is very deleterious. It will be found that the copper is not all precipitated by the cyanide of potassium, for according to this mode, when a precipitate ceases to be formed, the solution remains greenish blue, probably owing to the decomposition of the cyanate of potash, and the formation of ammonia, which holds copper in solution, and forms also some complicated compounds with the cyanides of copper. If cyanide of potassium is added until the blue solution disappears, still copper is held in the solution, and may be detected by taking out a little, and adding to it a few drops of sulphuric acid, which will give a white precipitate of subcyanide of copper. The loss of copper sustained is the only objection to this mode of preparing a copper solution. The cyanide of potassium is added until a precipitate is no longer formed; it is then allowed to settle, the clear liquid is poured off, and the vessel is to be filled with water: when the precipitate has again settled, the liquor is poured off, and this washing is repeated four or five times, in order to wash out the sulphate of potash which is formed during the precipitation. After being thus washed, a solution of cyanide of potassium is added to the precipitate until it dissolves. The coppering solution is now complete: it is of a light yellow color, and is well adapted for ordinary purposes. The loss of copper is, however, considerable, being about one-fifth of the whole.

*Second Method.*—A coppering solution may also be prepared by adding cyanide of potassium, to oxide of copper, or to carbonate of copper, until it is dissolved. But these solutions are objectionable, the latter especially so, as it contains a great quantity of carbonate of potash, formed from the mutual decomposition of the carbonate of copper and cyanide of potassium, and the carbonate of potash deteriorates the solution; the former leaves potash in the solution, but this is not so bad as the carbonate of potash.

*Third Method.*—The method we have adopted in manufacturing purposes is as follows:—To a solution of sulphate of copper, we add a solution of ferrocyanide of potassium, so long as a precipitate continues to be formed: this is allowed to settle, and the clear liquor being decanted, the vessel is filled with water, and when the precipitate settles, the liquor is again decanted, and we continue to repeat these washings until the sulphate of potash is washed quite out. This is known by adding a little chloride of barium to a small quantity of the washings, and when there is no white precipitate formed by this test, the precipitate is sufficiently washed. A solution of cyanide of potassium is now added to this precipitate until it is dissolved, during which process the solution becomes warm by the chemical reaction that takes place. The solution is filtered, and allowed to repose all night. If the solution of cyanide of potassium that is used is strong, the greater portion of the ferrocyanide of potassium crystallizes in the solution, and may be collected and preserved for use again. If the solution of cyanide of potassium used to dissolve the precipitate is dilute, it will be necessary to condense the liquor by evaporation, to obtain the yellow prussiate in crystals; the remaining solution is the coppering solution. Should it not be convenient to separate the yellow prussiate by crystallization, the presence of that salt in the solution does not deteriorate it, nor interfere with its power of depositing copper.

PECULIARITIES IN WORKING CYANIDE OF COPPER SOLUTION.—The true composition of the salts thus formed by copper and cyanide of potassium has not yet been determined, being both various and complicated, but their relations to the battery and electrolyzation are peculiar. The solution must be worked at a heat of not less than from 150° to 200° Fahrenheit. All other solutions we have tried follow the laws laid down by Spencer and Smee, namely, that if the electricity is so strong as to cause gas to be evolved at the electrode, the metal will be deposited in a sandy or powdered state; but the solution of cyanide of copper and potassium is an exception to these laws, and there is no reguline deposit obtained unless gas is freely evolved from the surface of the article upon which the deposit is taking place. This necessitates the use of batteries of several pairs intensity, varying from five to nine pairs of Wollaston's battery, according to the heat and the state of the solution.

As this solution is used hot, a considerable evaporation takes place, which requires that additions be made to the solution from time to time. If water alone is used for this purpose, it will precipitate a great quantity of copper as a white powder, but this is prevented by dissolving a little cyanide of potassium in the water at the rate of about four ounces to the gallon. The vessels used in factories for this solution are generally of copper, which are heated over a flue, or on a sand-bath—the vessel itself serving as the positive electrode of the battery; but any vessel will suit if a copper electrode is employed, when the vessel is not of copper.

PREPARATION OF IRON FOR COATING WITH COPPER.—When it is required to cover an iron article with copper, it is first steeped in hot caustic potash or soda, to remove any grease or oil. Being washed from that, it is placed for a short time in dilute sulphuric acid, consisting of about one part of acid to sixteen parts of water, which removes any oxide that may exist. It is then washed in water, and scoured with sand till the surface is perfectly clean, and finally attached to the battery, and immersed in the cyanide solution. All this must be done with despatch, so as to prevent the iron from combining with oxygen. An immersion of five minutes duration in the cyanide solution is sufficient to deposit upon the iron a film of copper. But it is necessary to the complete protection of the iron, that it should have a considerable thick coating: and, as the cyanide process is expensive, it is preferable, when the iron has received a film of copper by the cyanide solution, to take it out, wash it in water, and attach it to a single cell or weak battery, and put it into a solution of sulphate of copper. If there is any part not sufficiently covered with copper by the cyanide solution, the sulphate will make these parts of a dark color, which a touch of the finger will remove. When such is the case, the article must be taken out, scoured, and put again into the cyanide solution till perfectly covered. A little practice will render this very easy. The sulphate solution for covering iron should be prepared by adding to it by degrees a little caustic soda, so long as the precipitate formed is re-dissolved. This neutralizes a great portion of the sulphuric acid, and thus the iron is not so readily acted upon.

EFFECTS OF CONDUCTING POWER IN SOLUTIONS AND METALS.—In covering iron, platinum, or such comparatively bad-conducting metals, with other metals that are good conductors, or the solutions of which are good conductors, the property of conduction in relation to the solution is beautifully illustrated. If we take a copper wire, say 8 or 10 feet long, one end of which is attached to the zinc of a battery, and laid parallel with the positive electrode into a solution for the purpose of receiving a deposit, it will be found that the greatest amount of deposit has taken place at the end furthest from the battery: but if an iron or platinum wire be substituted for the copper one, the contrary result will take place; for the end furthest from the battery will be the last to receive the coating, and will have the least quantity of metal deposited upon it. If the copper wire was 30 feet long, little alteration would be seen in the deposit; but upon an iron or platinum wire of that length the deposit proceeds only a certain distance, and no deposit will take place on the end furthest from the battery until the current has passed a considerable time, after which the deposit is observed to advance gradually. The copper as it becomes deposited on the iron acts as a conductor, transmitting the deposit further onwards to its final point, as well as adding to the deposition already effected upon the iron. The length of deposit that would be formed on the first immersion of the wire depends upon the conducting power of the solution; for, as already stated, solutions vary in this property as well as metals. We have found that a few feet of iron wire offer a greater resistance to the passage of the current than the solution between the iron wire and the positive electrode, which is only about 2 or 3 inches; but their exact relations to each other we have not yet had an opportunity of investigating.

Under these circumstances, it may be asked, why not increase the intensity of the battery, and so force it along the wire? But this, as will be apparent, can only be done within certain limits; for by increasing the intensity of the battery it may be rendered too strong for the solution near the battery, and thus a sandy deposit will be given at the one end and none at the other. The electro-metallurgist, when coating long rods of iron wire with any metal, has to make connections with the battery every few feet. The wire is generally coiled up in the form of a cork-screw, and suspended by copper wires. We have found it very convenient to coil it upon a reel, having its armatures tipped with copper, and connected with the battery. This plan insures a regular coating, but the position of the wire requires to be changed during the operation, otherwise the parts which press upon the arms of the reel will be left without deposit.

Illustration of Conduction.—As an illustration of the property of conduction, we mention the following circumstance:—Having a large iron shaft, or rod, about 12 feet long and 3 inches average diameter, to cover with copper, we had it properly cleaned, placed in a hot solution of cyanide of copper and potassium, and surrounded by sheets of copper as a positive electrode. Two batteries of 7 pairs intensity were attached, one at each end of the shaft; but, by an oversight, one of the batteries was not properly connected, the copper terminal of the battery having been attached to the shaft. Had the shaft been of copper, the one battery would have neutralized the other, so that there would not have been any deposit; or, had the one battery been stronger than the other, there would have been a current and deposit equal to the excess of power of the one over the other. But, under the stated circumstances, a different result was obtained. After the batteries had been in action two hours, we found that a beautiful copper coating was imparted to that half of the shaft which extended from the point properly connected, while the other half was quite bare—no deposit having taken place upon it: but a deposit had been made upon the copper electrode opposite this non-affected half. The batteries did not (as we could perceive) affect one another, except that the one improperly connected prevented the deposit effected by the other proceeding further than the half length of the shaft; but it made the deposit obtained more perfect than would have been the case had there been only one battery at one end.

In this instance, the distance of the shaft from the electrode was 6 inches, so that the resistance of 6 feet of the iron was more than 6 inches of the solution; hence the influence of the contra-acting battery could not reach further: or if any power passed further it was neutralized by the other battery—which we are inclined to think did not take place—as the amount or thickness of deposit upon the one half was fully more than we would have anticipated upon the whole, had the batteries been properly connected.

NON-ADHERENCE OF DEPOSIT.—Objections have been made to covering iron with copper for its protection, from an impression that the copper will not adhere to the iron; but if the operation is carefully performed the copper will adhere; when it does not, it will generally be found that it is the copper deposited from the sulphate which loosens from the copper deposited from the cyanide—occasioned no doubt, by the article not having been sufficiently washed from the cyanide solution, and thus having a thin film of cyanide of copper precipitated upon the surface, which prevents the adhesion of the after deposit. Or, as it happens sometimes, that the cyanide of copper solution has not much free cyanide of potassium, and, consequently on putting the article into water, the cyanide of copper is decomposed by the water, and precipitated upon the surface. If a little cyanide of potassium is dissolved in the first water used for washing out the depositing solution, this will be prevented.

We have repeatedly deposited copper upon iron wire, and afterwards had it drawn out to twice its original length without the copper striping off; but, as the copper becomes hard and brittle, it is liable to break if the wire is much bent, and if it be made red hot, to anneal or soften it, the copper will oxidize, and if the coating is thin, the iron will be left bare in some places. We have seen iron bolts, covered with copper, driven through 17 inch wood, and nails of all sizes subjected to rough work, without the deposit being injured. Some iron work coated in 1842, and exposed to the atmosphere, remains in good condition still. These remarks are also applicable to iron covered with zinc. The coating of iron with copper has been tried in a great variety of ways for large operations, but in general these trials have, commercially, ended unsuccessfully; the labor and cost is greater than the advantages sought will warrant for ordinary purposes. Many years ago trials were made to cover cast-iron with copper, and then gild or plate for ornamental use, but only with partial success. More recently, however, a patent has been taken out for the same purpose, and which, being one of these applications that would be useful in beautifying and improving the taste of the community, we give an abstract of the specification as follows:

COATING CAST-IRON WITH OTHER METALS.—"Mr. W. Newton (for a correspondent) has patented the coating cast-iron permanently with copper, by depositing the copper by galvanic action, from a solution prepared by first taking a saturated solution of sulphate of copper in water and precipitating with carbonate of potash, and then re-dissolving in cyanide of potassium, whether the copper be deposited directly on the surface of the cast iron, or on zinc previously deposited thereon. The second part of this invention consists of coating cast-iron with the alloy of copper called brass, by first coating the cast-iron with copper, or zinc, or both, and then depositing the brass thereon, by galvanic action, from a solution formed by mixing with the solution of copper employed in the first part of the invention, a solution of zinc prepared in substantially the same manner. The iron articles thus coated, may be subsequently coated with gold or silver, so as to give them the appearance of these latter metals. The articles of cast-iron to be coated or plated are first to be cleansed by what is known as the 'pickling' process, with dilute sulphuric acid, and then 'scratch-brushed,' as it is termed, to free the surface from scale, sand, and other foreign substances which may not have been removed by the acid; and after this the castings are to be immersed in dilute nitro-muriatic acid. Any other mode of thoroughly cleansing the surface may be substituted for that above indicated. A solution of zinc is then prepared in the following manner:—Dissolve the sulphate of zinc in water until the water is saturated, and precipitate by means of prussiate of potash. The precipitate is then collected in a filter, and re-dissolved in cyanide of potassium. This constitutes solution number one. A solution of copper is then prepared in the same manner, by dissolving sulphate of copper in water, and precipitating with carbonate of potash: this precipitate is dissolved in cyanide of potassium, and is called the second or copper solution. The third, or what may be termed the brass solution, is then prepared by mixing together the first or zinc solution with the second or copper solution, in such proportions as to produce the shade of color required—increasing the proportional quantity of the one or the other at the discretion of the operator. The iron castings having been thoroughly cleansed, are first immersed in the first or zinc solution, and the galvanic battery applied in the usual manner of electrotyping and continued until the required thickness of zinc is deposited on the surface of and caused to unite with the surface of the cast-iron, which is a carbonate of iron. The castings thus coated or plated with zinc, are then to be immersed in the second or copper solution, and the galvanic battery applied, as with the first or zinc solution, and continued until the required thickness of copper shall have been deposited. In this way it will be found that the copper coating has become thoroughly attached to the zinc, and the zinc to the iron, so that they cannot be removed except by filing or cutting, as in the case of a solid mass of copper; so that articles, of whatever form desired, which can be made of cast-iron—that is, of carbonate of iron—can be coated with copper, so as to answer nearly if not all the purposes to which they could be applied if made of solid copper; thus greatly economizing the cost. After the surface of cast-iron has been coated with zinc, or with copper, or with zinc and then with copper, which latter is much the best, if it be desired to coat it with brass, it is to be immersed in the third or brass solution, and the galvanic battery applied, until the required thickness shall have been deposited. In doing this it is important that the positive pole of the battery should be made of brass, and as nearly as practicable of the shade of the brass to be deposited; for if a copper pole be applied, it will deposit in excess the copper portion of the solution. If desired, the brass can be deposited on the coating of zinc instead of the coating of copper; but it will be found decidedly better to deposit the brass on the coating of copper, whether the copper be deposited directly on the cast-iron or on a coating of zinc, although the latter is the best. In this way articles are produced, having all the appearance, and answering nearly if not all the same purposes as if made entirely of brass, and at much less cost. The cast-iron being thus coated with brass, the surface may be bronzed in the usual and well-known manner of bronzing brass; and as the process of bronzing on brass and copper is well known, it will be unnecessary to give a detailed description of it. The surface of the cast-iron being thus coated with brass, or with copper, can then be coated effectually with silver or gold in any of the well known modes of coating brass or copper with those fine metals; it will not, however, be necessary to give the details of such mode or modes, as they are well known in the arts. The patentee remarks that it will be found better to deposit the silver or gold on the brass coating than on the copper coating, on account of the color—particularly when, from reasons of economy, it is desired to make the coating of fine metal very thin. The patentee claims the process herein described, or any mere modification thereof, for coating cast-iron (carbonate of iron) with copper, by causing the copper, from a solution such as above described, to deposit, by galvanic action, directly on the surface of the cast-iron, or on the zinc previously deposited thereon, as set forth. And also the process herein described, or any mere modification thereof, for coating cast-iron with the alloy of copper, known as brass, by causing the brass, from a solution such as above described, to deposit, by galvanic action, on to the surface of the cast-iron, previously coated with zinc, or copper, or both, as specified."

Coating of Iron with Zinc.—In covering iron with zinc, the precautions necessary for copper are not required: zinc being the positive metal, acids have a stronger affinity for it than for iron, and therefore an acid solution may be used. The one generally used is the sulphate.

Sulphate of Zinc.—Zinc dissolves easily in sulphuric acid, and the solution by evaporation yields crystals of sulphate of zinc; but as the salt is very cheap and abundant in the market, it is more convenient and economical to buy than to make it. The solution for depositing is made by dissolving 2 lbs of the crystallized salt in one gallon of water. The single cell process cannot be used advantageously with this solution. A separate battery is necessary, and a zinc positive electrode. The metal is very easily deposited —one or two pairs of Wollaston's battery being sufficient for coating small articles.

Zinc may be deposited upon black-leaded surfaces in the same manner as copper; but, unless more than ordinary precautions are observed, an article formed in this manner is so brittle that it can hardly be handled without breaking, from its crystalline character. When the deposition upon black-lead is attempted, the best method is to have the solution saturated with salt, employing a battery of six or seven pairs of plates, and keeping the articles on which the deposit is taking place constantly in motion.

The use of cyanide of zinc has been recommended, but for what good reason it is hard to know. It is unnecessary, and its use presents great practical difficulties. The positive electrode becomes coated, after a few minutes working, with a white pasty matter which prevents further action and stops the current. Some of this white coating collected, washed, and dried in the air, gave by analysis,

| | |
|---|---|
| Oxide of zinc . . . . . . | 51·3 |
| Cyanogen . . . . . . . . . | 1·7 |
| Iron . . . . . . . . . . . | trace |
| Potash . . . . . . . . . . | 2·3 |
| Carbonic acid . . . . . . | 27·8 |
| Matter insoluble in HCl. . . . | 2·5 |
| Water . . . . . . . . . . | 14·8 |
| | 100·4 |

The zinc is converted into carbonate of zinc: the potash is combined with the cyanogen as cyanide of potassium.

Use of Zinc Coating.—The principal application of zinc is upon iron, to protect it from corrosion, which it does, not only as a coating, but, from its more electro-positive character, it protects it by a galvanic influence. The voltaic influence of zinc for protecting iron is a subject that has occupied the attention of practical men for a long time; it is one of high importance: nevertheless there seems yet a great deficiency in our knowledge of the extent of this influence, and how and when it is effective.

Upon this subject, Professor Faraday, in the Report of the Harbors of Refuge Commissioners, states, "Zinced iron would no doubt resist the action of sea-water so long as the surface was covered with zinc, or even when partially denuded of that metal: but zinc dissolves rapidly in sea-water, and after it is gone, the iron would follow.

"As to voltaic protection, it has often struck me that the cast-iron piles proposed for lighthouses, or beacons, might be protected by zinc in the same manner as Davy proposed to protect copper by iron; but there is no doubt the corrosion of the zinc would be rapid. If not found too expensive, the object would be to apply the zinc protectors in a place where they could be examined often, and replaced when rendered ineffective. In this manner I have little doubt that iron would be protected in sea-water."

INFLUENCE OF GALVANISM IN PROTECTING METALS FROM DESTRUCTION BY OXIDATION AND SOLUTION.—The galvanic influence of one metal in protecting another is in relation to their negative and positive qualities together with their conducting powers (p. 515). Their relations in sea-water are—silver, copper, bismuth, antimony, iron, tin, lead, cadmium, zinc; the first the most negative, the last the most positive in the series. So that, according to this scale, the further apart the metals may be which are selected for experiment, the more decided will be the power of the positive to protect the negative. Copper and zinc operate more strongly together than iron and zinc.

A metal that is insoluble when placed singly in a fluid, may be made soluble by connection with a relatively negative metal placed in the same fluid. For example, pure zinc put into muriatic acid is unaffected, but when connected with copper in the same fluid it is rapidly operated upon. Or a metal may be soluble in a fluid alone, but may be rendered insoluble by connection with a relatively positive metal which undergoes decomposition instead. Thus: copper is dissolved in sea-water when alone, but when a piece of zinc is connected with it, the copper is unaffected. This last effect is the substance of Davy's method of protection alluded to by Dr. Faraday, in applying the principle of which it is necessary to take into consideration,

1st, The amount and power of electricity generated by the connected metals in the same fluid; and

2d, The conducting power of the metal which is being protected.

1st. The amount and power of the electricity evolved is in proportion to the difference of the relative negative and positive condition of the metals employed. The more negative the coated metal is, the less it requires protection, although its powers of protection are the greatest. And the more positive the coated metal, the more liable it is to be destroyed, and the greater the amount of electricity required to protect it; but unfortunately it is less able to generate this electricity when in contact with another metal. Thus these two conditions are opposed to the application of galvanic influence for protecting iron.

Suppose, for example, that 4 square inches of zinc, in connection with 4 square feet of copper, give out sufficient electricity to protect the copper from sea-water, it will be found that to obtain the same amount of electricity by iron and zinc, 2 square feet of the latter to 4 square feet of the former are required.* Besides which, the same quantity of electricity that protects copper will not pro-

---

* These proportions, given in round numbers, are nearly accurate; but they vary according to the kind of iron, the state of the water, the distance of the metals, etc.

tect iron; nor will any quantity of zinc protect iron from corrosion in sea-water—even a bar of iron placed in a zinc vessel filled with sea-water is not completely protected.

2d. The conducting power of the negative or protected metal subjected to submarine immersion is a subject of very great importance. Suppose a piece of copper and a piece of zinc be connected under a solution—say a copper bar (c) 4 feet long, with a piece of zinc (z) 4 inches in length, erected on one end, as in the annexed sketch:

Fig. 580.

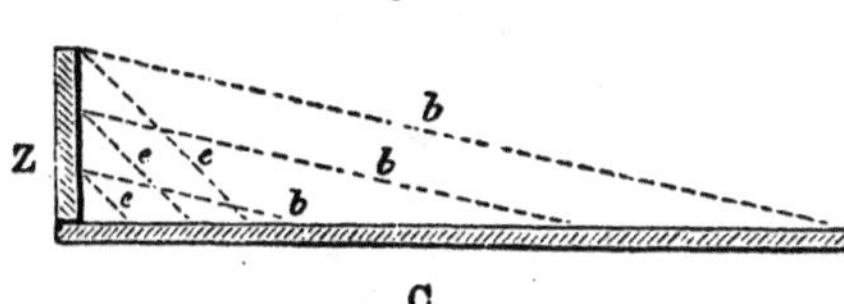


The conducting power of the copper is so much superior to that of the solution that the whole length of the bar will become instantly negative, and the current of electricity will pass to and from all parts of the bar at the same time in the lines *b*, *b*, *b*; but the current will be more active towards the point of contact than towards the distant extremity—the resistance of the solution being less in proportion to the proximity of the metals. But if a bar of iron, and a piece of zinc as a protector, be placed in the same circumstances, the phenomena assume quite a different aspect: the conducting power of iron being much less than that of copper, the distant extremity will not be affected by the electric current, which will find a more easy passage, as indicated by the dotted lines *e*, *e*, *e*, beyond which the electric effort ceases; and even in that portion of the bar which is under the influence of the current, the part nearest the zinc is better defended than those parts which are farther distant. This partial protection, while it induces a negative state at the near end, renders the other end more positive. Such a diversity of condition gives rise to voltaic action between the two extremities of the bar, and the result is the destruction of the far end. In all cases of voltaic protection the more equal the influence over the whole surface protected, the more perfect is the protection. An inequality of protection, such as we have described, is productive of numerous evils. It is, we believe, the source of many of the injuries occurring in our day to copper sheathing. One part of a sheet becoming, by some local cause, negative, the other parts are thus rendered positive; the result is, that upon the borders of an individual sheet either overlapping or underlying its neighboring sheet, an electric current is excited, passing through the stratum of moisture which may intervene, and the ultimate effect is that the positive edge is dissolved as effectually as if cut by a knife. The evil arising in one place may be so contagious as to affect a whole neighborhood—sometimes the whole side of a ship's bottom.

In fresh water iron cannot be protected any length of time, for the zinc coating speedily passes into a blackish substance, which peels off and exposes the iron to rust. When iron is simply exposed to the air, a good coating of zinc is a sure protection. We have seen iron of various qualities coated by the electro-processes and exposed to the atmosphere, in all weathers, for several years, without being more affected than a piece of zinc would be. In spots where abrasion has taken place by accident, the protecting power of the zinc is lost, and the iron rusts as if there were no zinc present. No other result, however, could be anticipated, as there can be no electric excitation without a liquid to connect the two metals.

The iron to be coated by zinc is to be cleaned and prepared in the same manner as we have described for the purpose of covering it with copper (page 579).

---

## CHAPTER XXX.

### ELECTRO-PLATING.

THE next applications of the electro-deposition we have to notice are those relating to silver and gold, embracing the arts of electro-plating and gilding—arts which are gradually revolutionizing some extensive branches of manufacture, having the same object but acting by a different means.

TO MAKE SILVER SOLUTION.—The solution of silver used for plating consists of cyanide of silver dissolved in cyanide of potassium which may be prepared in various ways. We shall first describe some of the preparations most in use, and also point out practical objections which, in special cases, have occurred under our own observation, not omitting to specify and recommend those methods which have approved themselves to us as being most simple and effective.

The method generally adopted is as follows:—Metallic silver is dissolved in four parts of nitric acid, diluted with one part of water: the diluted acid is heated in a vessel, and the silver is added by degrees. The operator must avoid breathing the fumes which ascend, as they are highly deleterious. The metal being dissolved, the solution is transferred to a large vessel, and diluted with water. To this is added a solution of cyanide of potassium so long as a white precipitate is formed. This precipitate is cyanide of silver, and the action which ensues may be thus represented:

| *Substances used:* | *Substances produced:* |
|---|---|
| Nitrate of silver $= AgO, NO^5$ | Nitrate of potash $= KO, NO^5$ |
| Cyanide of potassium $= KCy$ | Cyanide of silver $= AgCy$ |

$$AgO, NO^5 + KCy = KO, NO^5 + AgCy$$

The propriety of diluting the nitrate of silver before precipitating by the cyanide of potassium arises from the fact, that the salts of potash and soda (such as the nitrates, chlorides and sulphates), when in strong solution, dissolve small quantities of the silver salt, and thus cause a loss, which is prevented by previous dilution with water.

When the precipitate of cyanide of silver has settled, the clear solution is carefully decanted, and the vessel filled up with water, which is again decanted as soon as the precipitate has settled. This process is to be repeated three or four times, so as effectually to wash out the soluble salts. When properly washed, a solution of cyanide of potassium is added to the precipitate, until it is all dissolved. The resulting solution constitutes the cyanide of potassium and silver, and forms the plating solution. It ought to be filtered previous to using, as there is always formed a black sediment, composed of iron, silver, and cyanogen, which, if left in the solution, would fall upon the surface of the article receiving the deposit, and make it rough. The sediment, however, must not be thrown away, as it contains silver. The cyanide of potassium, used to dissolve the cyanide of silver, may be so diluted that the plating solution, when formed, shall contain one ounce of silver in the gallon: of course the proportion of silver may be larger or smaller, but that given is what we consider best for plating.

In dissolving 100 ounces of silver, the following proportions of each ingredient are those which we have found in practice to be the best. Take 7 pounds of the best nitric acid* and 61 ounces of cyanide of potassium, of the average quality described at page 576; this quantity will precipitate the 100 ounces of silver dissolved in the acid solution. After this is washed, take 62 ounces more of cyanide of potassium, the solution of which will dissolve the precipitate: this being done, the plating solution is then formed. Of course, these proportions will vary according to the difference in the quality of the materials; but they will serve to give an idea of the cost of the silver solution prepared in this manner.

CYANIDE OF SILVER DISSOLVED IN YELLOW PRUSSIATE OF POTASH.—We have occasionally dissolved the cyanide of silver by yellow prussiate of potash, three pounds of which are required to dissolve one ounce of silver. This forms an excellent plating solution, and yields a beautiful surface of silver. It must have a weak battery power, and consequently the silver is very soft. The positive electrode does not dissolve in this solution: there is formed upon its surface a white scaly crust, which drops off and falls to the bottom; and the solution soon becomes exhausted of silver.

SOLUTION MADE WITH OXIDE OF SILVER.—It has been recommended to dissolve the oxide of silver in cyanide of potassium, which forms a solution of cyanide of potassium and silver; but this preparation is less economical, because the materials used in converting the silver into an oxide are lost: it requires the same amount

* The nitric acid must be free from hydrochloric (muriatic) acid: to a small quantity of the acid add a few drops of solution of nitrate of silver; if it gives a milky white precipitate, it contains muriatic acid, and should be rejected.

of cyanide of potassium as the process just described, and brings, moreover, an equivalent of potash into the solution, which is a disadvantage. The following diagram shows the reactions that occur:

| *Substances used:* | | *Substances produced:* | |
|---|---|---|---|
| Oxide of silver | = AgO | Potash = KO | |
| Cyanide of potassium | = KCy | Cyanide of silver and potassium | =KCy, AgCy |
| Cyanide of potassium | = KCy | | |

AgO+2 KCy=KO+KC, AgCy.

Solution made with Chloride of Silver.—The nitrate of silver may also be precipitated by adding a solution of common salt to it, and treating it in the same way as described for precipitation by cyanide of potassium: this would form chloride of silver, which may be dissolved in cyanide of potassium, thus forming the silver solution. But the objection urged against the use of oxide of silver is equally applicable in the case of chloride; and much greater care is required in precipitating large quantities and strong solutions of silver by common salt, than by cyanide of potassium, the chloride of silver being more soluble in the salts of the alkalies—as the nitrates, chlorides, and sulphates—than cyanide of silver is; and there is therefore great liability to loss by this process, in which we have not the redeeming quality of a saving of materials, as the following diagram will show:

| *Substances used:* | | *Substances produced:* | |
|---|---|---|---|
| Chloride of silver | = AgCl | Chloride of potassium | = KCl |
| Cyanide of potassium | = KCy | Cyanide of silver and potassium | =KCyAgCy |
| Oyanide of potassium | = KCy | | |

AgCl+2 KCy = KCl+KCy, AgCy.

Thus, we observe, that the action taking place is not mere solution, but decomposition; which upon one hundred ounces of silver in this preparation produces an impurity of seventy ounces of chloride of potassium, which, although not very injurious to the solution, would be much better away.

The Best Method of Making Silver Solution.—The best and cheapest method of making up the silver solution is by the battery, which saves all expense of acids and the labor of precipitation. This is effected by taking advantage of the principle of non-transfer of metal in electrolytes (see page 561). To prepare a silver solution which is intended to have an ounce of silver to the gallon (see p. 588), observe the following directions: Dissolve 123 ounces of cyanide of potassium in 100 gallons of water; get one or two flat porous vessels, and place them in this solution to within half an inch of the mouth, and fill them to the same height with the solution; in these porous vessels place small plates or sheets of iron or copper, and connect them with the zinc terminal of a battery: in the large solution place a sheet or sheets of silver connected with the copper terminal of the battery. This arrangement being made at night, and the power employed being two of Wollaston's batteries, of five pairs of plates, the zincs 7 inches square, it will be found in the morning that there will be dissolved from 60 to 80 ounces of silver from the sheets. The solution is now ready for use: and by observing that the articles to be plated have less surface than the silver plate forming the positive electrode, for the first two days, the solution will then have the proper quantity of silver in it. We have occasionally found a little silver in the porous cell: it is therefore not advisable to throw away the solution in them without first testing it for silver, which is done by adding a little muriatic acid to it.

The amateur electrotypist may, from this description, make up a small quantity of solution for silvering his medals or figures. For example, a half-ounce of silver to the gallon of solution will do very well; a small quantity may be prepared in little more than an hour.

As the cyanide of potassium dissolves silver without the aid of a battery, by merely allowing a piece of silver to steep in this solution for a few days, a plating liquor may be formed; but this is tedious and uncertain, although for small operations, and where porous vessels are not convenient, it will serve the purpose.

Other solutions of silver may be employed if the law stated at page 575 is strictly observed. Indeed, every salt of silver has not only been tried, but is either the subject of a patent, or prominently included in it. None of them, however, with the exception of two, have we found of any practical value, besides that already described: these are the chloride of silver dissolved in hyposulphite of soda, and the sulphite of silver dissolved in sulphite of potash or sulphite of soda.

Hyposulphite of Silver Solution.—The simplest method known to us for forming the hyposulphite of silver solution is this: Take one pound of pure carbonate of soda, well dried, as described at page 576; mix it intimately with five ounces of flour of sulphur; place the mixture over a slow fire without flame in a porcelain or stoneware basin, which must be supported by an iron trelis, or any convenient support, to prevent it touching the red coal or flame; keep the mixture constantly stirring, and maintain the heat till the sulphur melts, and the mass inclines to get pasty and rough. While in this state keep stirring for about fifteen minutes, in order to bring every part in contact with the air. Set the mixture to cool—after which dissolve in water. Boil the solution for some time, adding sulphur; then filter it, and allow it to evaporate at a slow heat. The crystals formed are hyposulphite of soda.

To prepare the silver solution, the silver is first dissolved in nitric acid, and then precipitated by a solution of common salt, and washed, with the precautions stated at page 587. When the precipitated chloride of silver is well washed, some of the crystals of hyposulphite of soda are dissolved, and the solution is added to the chloride of silver, which it dissolves, forming the plating solution. It is not necessary to crystallize the hyposulphite of soda, if used as soon as made.

The hyposulphite of silver solution is very easily decomposed by the electric current, so that a weak battery will suffice to plate by it: but its great objection is its liability to decompose in the light, and to deposit the silver as sulphuret: unless great care is exercised, the silver deposited from it will be in a granular condition, which is a great objection in plating.

SULPHITE OF SILVER PLATING SOLUTION.—The sulphite of silver solution is prepared in the following manner, as described by the patentee of the process:

"The solution which I use is made in the following manner: I take of the best pearl-ash of commerce 28 lbs (avoirdupois) and add to it 30 lbs (avoirdupois) of water, and boil them in an iron vessel until the pearl-ash is dissolved; the solution should then be poured into an earthenware or other suitable vessel, and suffered to stand until the liquor becomes cold. It should then be filtered, and 14 lbs (avoirdupois) of distilled water added thereto; sulphurous acid gas (obtained by any of the known processes) should then be passed into the filtered liquor until it is saturated, taking care not to add sulphurous acid gas in excess. The liquor should be again filtered, and the liquor so filtered is what I term the solvent, or sulphite of potash.

"To make the silvering liquor which I use in coating with silver the surface of articles formed of metal or metallic alloys, I dissolve 12 oz. (avoirdupois) of crystallized nitrate of silver in 3 lbs. of distilled water (in a clean earthenware vessel), and add to the solution, by a little at a time, the before-mentioned solvent, so long as a whitish colored precipitate is produced (care being taken not to add more of the solvent than is necessary). After the precipitate has subsided, I pour off the supernatant liquor, and wash the precipitate with distilled water. To the precipitate I add as much of the before-mentioned solvent as will dissolve it, and afterwards add about ⅛th part more of the solvent, so that the solvent may be in excess; I then stir them well together and let them remain about 24 hours, and then filter the solution, when it will be ready for use. This is what I designate silvering liquor."*

Fig. 581.

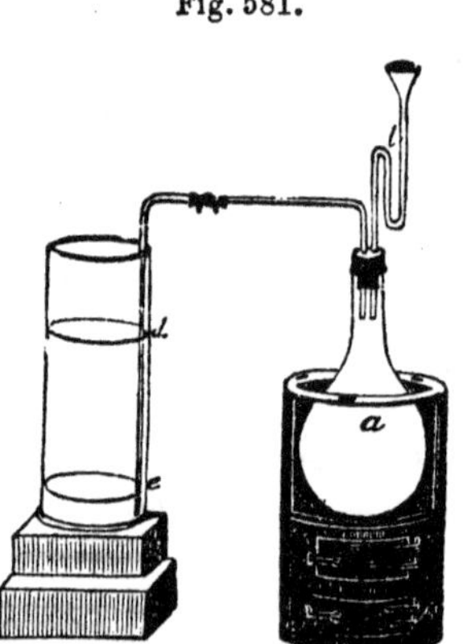


Sulphurous acid gas for making the above liquor may be pre-

* Repertory of Patent Inventions, 5th series, p. 210, 1843.

pared by heating sulphuric acid, undiluted, in a flask or any convenient vessel, to which should be added small pieces of copper or charcoal: the gas escaping is made to pass into the solution to be saturated with it.

Fig. 581 is a very convenient apparatus for the preparation of this gas for saturating solutions.

This solution is also very easily decomposed by the electric current, and serves the purposes of plating very well; but it is also liable to decomposition by light, and is not so good in practice as the solution of cyanide of potassium and silver. The latter solution is, however, liable to a kind of decomposition not yet fully investigated, but it is wholly confined to its impurities, and it never deposits its silver; whereas the decomposition that takes place in the sulphite or hyposulphite affects the silver compound, and precipitates the silver from solution.

TO RECOVER SILVER FROM SOLUTION.—When a silver solution gets out of order, and cannot be rendered fit for use again, the silver may be recovered by adding to the solution any acid that will neutralize the alkali; if nitric or sulphuric acid be used, the silver precipitates as cyanide, but if hydrochloric acid be used, the silver will be precipitated as a chloride: in either case the solution should be diluted, or a portion of the precipitate will be re-dissolved. The precipitate is allowed to deposit, the clear liquor decanted, and the vessel filled with water to wash the precipitate, which is afterwards collected upon a filter and dried, and then mixed with twice its weight of carbonate of potash, and fused in a Hessian crucible for 15 minutes, or until the fused fluid ceases to effervesce. On removing the crucible, and pouring the whole into an iron ladle, when cool the silver will be found in the metallic state at the bottom of the ladle.

In these operations when pouring the acid into the cyanide solution, great care must be taken not to inhale the fumes given off, which are very abundant, sickening, and poisonous. The operation should be done in the open air, and even then it is bad. Instead of throwing down the silver by an acid it is better to evaporate the solution to dryness, and to fuse the product as described; in which case the cyanide is an excellent reducing flux, requiring no addition of carbonate of potash, and saves the necessity of evolving poisonous fumes.

When the solution has contained yellow prussiate of potash, it is found that during this fusion portions of the metal sometimes form a scoriaceous nodule at the bottom of the crucible, and all the heat that can be applied by an ordinary assay furnace will not fuse it. This refractory piece, when cooled, has generally a rough scoriaceous surface, and is exceedingly hard. When filed it has more the color of German silver than of real silver; it has considerable malleability, and retains its bright appearance for a long time without tarnish. An analysis of this alloy gave—

| | |
|---|---|
| Silver . . . . . . . . | 82·15 |
| Copper . . . . . . . . | 9·12 |
| Iron . . . . . . . . . | 7·50 |
| Carbonaceous matter . . | ·46 |
| | 99·23 |

If we suppose the carbonaceous matter to be an accidental impurity, this alloy will nearly agree with the formula $Ag^3$ Cu Fe.

PREPARATION OF ARTICLES FOR PLATING.—Articles that are to be plated are first boiled in an alkaline ley, to free them from grease, then washed from the ley, and dipped into dilute nitric acid, which removes any oxide that may be formed upon the surface; they are afterwards brushed over with a hard brush and sand, of which a kind obtained from the Isle of Wight, and known as silver-sand is best. The alkaline ley should be in a caustic state, which is easily effected by boiling the carbonated alkali with slaked lime, until, on the addition of a little acid to a small drop of the solution, no effervescence occurs. The lime is then allowed to settle, and the clear liquor is fit for use. The ley should have about half-a-pound of soda-ash, or pearl ash, to the gallon of water. The nitric acid, into which the article is dipped, may be diluted to such an extent that it will merely act upon the metal. Any old acid will do for this purpose. In large factories the acid used for dipping before plating is generally afterwards employed for the above purpose of cleaning.

The article being thoroughly cleaned and dried, has a copper wire attached to it, either by twisting it round the article or putting it through any open part of it, to maintain it in suspension. It is then dipped into nitric acid as quickly as possible, and washed through water, and then immersed in the silver solution, suspending it by the wire which crosses the mouth of the vessel from the zinc of the battery. The nitric acid generally used and found best for dipping has a specific gravity 1·518, contains 10 per cent. sulphuric acid, and is got at about 6 cts. per lb. The article is instantaneously coated with silver, and ought to be taken out after a few seconds and well brushed. On a large scale, brushes of brass wire attached to a lath are used for this purpose; but a hard hair brush with a little fine sand will do for small work. This brushing is used in case any particle of foreign matter may be still on the surface. It is then replaced in the solution, and in the course of a few hours a coating of the thickness of tissue-paper is deposited on it, having the beautiful matted appearance of dead silver. If it is desired to preserve the surface in this condition, the article must be taken out, care being taken not to touch it by the hand, and immersed in boiling distilled water for a few minutes. On being withdrawn, sufficient heat has been imparted to the metal to dry it instantly. If it is a medal, it ought to be put in an air-tight frame immediately, or if a figure, it may be at once placed under a glass shade, as a very few days exposure to the air tarnishes it, by the formation of sulphuret of silver, and that more especially in a room where there is fire or gas. If the article is not wanted to have a *dead* surface, it is brushed with a wire brush, and old ale, beer, or water containing in solution a little gum, glue, or sugar, but the amateur may use a hard hair brush. It may be afterwards burnished according to the usual method of burnishing, by rubbing the surface with considerable pressure with polished steel or the mineral termed *bloodstone.*

We may remark, that in depositing silver from the solution, a weak battery may be used; though when the battery is weak the silver deposited is soft, but if used as strong as the solution will allow, say 8 or 9 pairs, the silver will be equal in hardness to rolled or hammered silver. If the battery is stronger than the solution will stand, or the article very small compared to the size of the plate of silver forming the positive electrode, the silver will be deposited as a powder. The average cost of depositing silver in this way is 4 cts. per ounce. Gas should never be seen escaping from either pole: and the surface of the article should always correspond as nearly as possible with that of the positive electrode, otherwise the deposit runs the risk of not being good; it requires more care, and the solution is apt to be altered in strength, because if the positive electrode be large compared with the negative, the solution will become stronger in silver, while if smaller in proportion the solution will become exhausted of silver.

In plating large articles (such as those plated in factories), it is not always sufficient to dip them in nitric acid; wash and immerse them in the solution, in order to effect a perfect adhesion of the two metals. To secure this, a small portion of quicksilver is dissolved in nitric acid, and a little of this solution is added to water, in sufficient quantity to enable it to give a white silvery tint to a piece of copper when dipped into it: the article then, whether made of copper, brass, or German silver, is, after being dipped in the nitric acid and washed, dipped into the nitrate of mercury solution till the surface is white: it is then well washed by plunging it into two separate vessels containing clean water, and finally put into the plating solution. This secures perfect adhesion of the metals. One ounce of quicksilver thus dissolved will do for a long time, though the liquor is used every day. When the mercury in this solution is exhausted, it is liable to turn the article black upon being dipped into it: this must be avoided, as in that case it also causes the deposited metal to strip off.

Practical Instructions in Plating.—We need hardly add that it is necessary that the battery should be so arranged, that the *quantity* of electricity generated should correspond with the surface of the articles to be coated, and that the *intensity* should bear reference to the state of the solutions; that is to say, that the *quantity* should be sufficient to give the required coating of metal in a given time, and the *intensity* such as to cause the electricity to pass through the solution to the articles. It is also essential for regular working, as stated above, that the plates of metal forming the positive pole in the solution should be of corresponding surface to the articles to be coated, and face them on both sides.

The following is the arrangement adopted in some of the large plating manufactories:—The *vat,* or plating vessel measures about 6½ feet in length, by 33 inches in breadth, and 33 inches in depth, and generally contains from 200 to 250 gallons of solution; the silver plates serving as electrodes, which were formerly nailed upon frames of wood, are now generally fixed upon light iron frames, these not being affected by the solution: two battery troughs are arranged as seen in Fig. 582, consisting of 6 batteries of three pairs intensity. The zinc plates immersed in the acid measure 6 inches by 7 inches, the exposed surfaces of which measure 84 square inches: these multiplied by 6 give 504 square inches, from which electricity is disengaged. The surface of the silver electrodes exposed to the articles receiving the deposit vary from 3000 to 4000 square inches of surface.

The following figure, with explanation, will illustrate these observations:

Fig. 582.

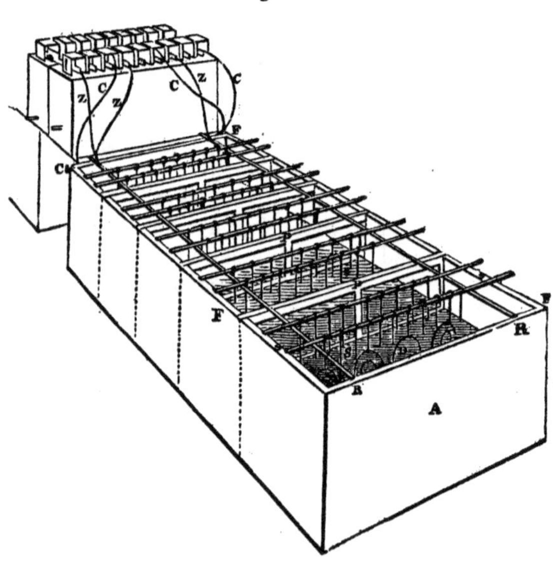


A, Vat or vessel containing the solution; B, Battery with zinc pole Z, connected with rods R R; and copper pole C, connected with the metallic sheets P P, in the solution, by means of the copper slip F; D D, are articles suspended in the solution by wires from the rods R R; S, the solution. So soon as the articles, which are connected with the negative pole of the battery, and the metallic sheets, connected with the positive pole, are both immersed in the solution, the galvanic circuit is completed.

In most plating establishments the batteries are now placed outside the house, and the connecting rods are brought from them into the vats, so as to preserve the workmen from the injury arising from inhaling the hydrogen gas which is given off by the zincs, as it often contains arsenic, and hence is highly injurious to health; the gas of the battery has a strong effect upon the nostril, exciting dryness with pain. Wherever the batteries are placed, they should not be exposed to cold, as their operation is much affected by the temperature.

In the early days of electro-plating the batteries used were round. They consisted of a copper cylindrical vessel, about 20 inches deep, and 5 inches diameter, filled with dilute sulphuric acid. A piece of wood was placed at the bottom of this vessel, and a cylinder of zinc, the same depth as the copper vessel, and about 3 inches diameter, was placed inside the copper vessel. A wooden ring, floating on the surface of the acid, prevented the zinc and copper touching—a binding screw was attached to each, and formed a battery of a single pair. Six batteries of this size were connected with such a vat as is described above. The test of strength employed to determine whether the working power was sufficient, was that, when connected with an electro-magnet, it should support a 7 lb. weight. We believe that many platers and gilders still use such batteries, and that, when the solutions and apparatus are all in good condition, they do well. They are, however, far from being so economical as the battery with square plates shown above. Some electro-metallurgists use large and deep stoneware vessels in which are placed the zinc and copper—the plates having several square feet of surface. We have already shown, when treating of batteries, that very large plates are not consistent with economy.

To ascertain the amount of metal deposited, it is only necessary to weigh the articles carefully before and after plating. But between the first weighing and the immersion of the articles in the plating solution there is the dipping into nitric acid to be accounted for: this, on an average, will cause a loss of about one pennyweight upon an article of the size of a foot square; thus, if a waiter of a foot square, made of copper or German silver, shows, when coated, a difference in weight of 19 pennyweights, the silver laid on must be estimated at an ounce, or 20 pennyweights. When the article is a "replate," *i. e.*, an old plated article that has become bare of silver in parts, the allowance or reduction for the dipping in the acid is only to include the portions left bare, for the silvered parts are not acted upon by it. One of the practical difficulties which the inexperienced will occasionally meet with when a "replate" is dipped in the nitric acid, is, that a galvanic action is produced between the silver and the copper portions, which causes a black line round the edge of the silver: this ought immediately to be rubbed off, but even with rapid and careful rubbing there is great danger that the coating will loosen and blister at those parts; and beside this, it happens that the parts of the "replate" which are sound, the silver not being acted upon by the acid, but rather protected by the galvanic action, are not in a fit state to receive and maintain a perfect adhesion of the deposit, and therefore the risk is great that the new coating will separate from the old, or, in technical language, that the part will *strip*. Under these circumstances experience has taught that the best way to proceed is to take all the old silver off the article, and deposit an entirely new coating.

There are two methods of taking off the silver:

TAKING SILVER FROM COPPER, ETC.—First, stripping or dissolving it off; this is done by putting into a stoneware or copper pan some strong sulphuric acid (vitriol), to which a little nitrate of potash is added: the article is laid into this solution, which will dissolve the silver without materially affecting the copper; saltpetre is added by degrees, as occasion requires; and if the action is slow a little heat is applied to the vessel. The silver being removed, the article is washed well and then passed through the potash solution, and finished for plating. When the sulphuric acid becomes saturated with silver it is diluted, and the silver is precipitated by a solution of common salt: the chloride of silver formed is collected and fused in a crucible with carbonate of potash, when the silver is obtained in a metallic state, as a knob or button. The crucible should not be over two-thirds full and should be kept in fusion till effervescence ceases. The crucible is then removed from the fire, and, when cool, it is broken.

The article thus stripped by acids often shows a little roughness, not from the effects of the acid, but because the copper under the silver has not been polished; it is therefore a necessary practice in the electro-plating factories to polish the articles before plating. This is done by means of a circular brush, more or less hard as required, fixed upon a lath, and a thin paste made of oil and pumice-stone ground as fine as flour. By this process the surface of any article can be smoothened and polished; but a little experience is required to ensure success, and enable the operator to polish the surface equally without leaving brush marks. We need scarcely say, that after this process the article must be cleaned in potash before it is plated.

*Second Method.*—Instead of stripping off the silver by means of acid, it is a more common and preferable mode to brush off the silver by the operation just described. In this case the brushings must be collected, dried, and burned; this may be done in an iron pan, keeping it at a red heat until all carbonaceous matters are consumed, the remainder is fused with carbonate of soda or potash, when the silver is obtained, in combination with a little copper.

CYANIDE OF SILVER AND POTASSIUM, ITS DECOMPOSITION DURING THE PLATING PROCESS.—The silver salt in the plating solution is a true double salt, being, as already described, a compound of one equivalent of cyanide of silver, and one of cyanide of potassium

—two distinct salts. In the decomposition of the silver solution by the electric current, the former, cyanide of silver, is alone affected: the silver is deposited, and the cyanogen passes to the positive plate or electrode. The cyanide of potassium is therefore set at liberty upon the surface of the article receiving the silver deposition, and its solution being specifically lighter than the general mass of the plating solution, rises to the top: this causes a current to take place along the face of the article being plated. If the article has a flat surface, suppose that of a waiter or tray, upon which a prominence exists, as a mounting round the edge, such as a gadroon, see Fig. 583, it will cause lines and ridges from the bottom to the top, as already described at page 561. Newly-formed solutions are most subject to produce this annoyance.

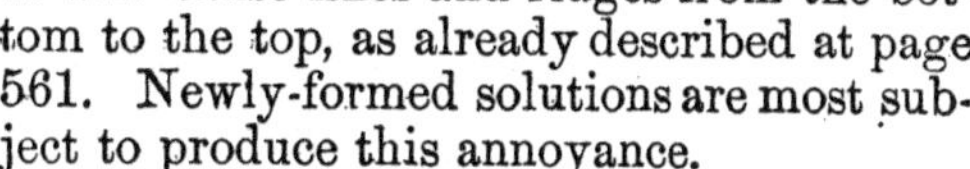

Fig. 583.

OTHER EFFECTS PRODUCED IN WORKING.—As the cyanogen combines with the silver plate forming the positive electrode, it is dissolved by the free cyanide of potassium, which the solution must have; and, being specifically heavier, sinks to the bottom, by which a current downwards is excited: this is of no greater annoyance than that it renders the solution of unequal density, which in its turn yields an unequal deposit, more being laid upon the lower parts of the article than on the upper: the silver plate also is destroyed more rapidly at the bottom than at the top, except at the surface of the solution, if the silver be above it, where the plate gets cut through. In a new solution, which contained 1½ ounces of silver to the gallon, we have found, just before taking out the articles, that the top part of the solution contained 200 grains of silver less, and the bottom part 200 grains more per gallon, than when the articles were put into it. These difficulties and annoyances may, however, be nearly surmounted by keeping the articles in motion: agitating or stirring the solution occasionally would also obviate these annoyances; but this is not advisable, for if the sediment (which always forms) were stirred up it would settle upon the face of the articles and make them rough. Where there is engine power it is an easy matter to keep the articles in motion; but where this power is not available, a very simple apparatus, invented by Mr. Alexander Mitchell, of Glasgow, may be fitted up at a trifling cost, to give the necessary motion by clockwork. The annexed sketch exhibits this apparatus.

MACHINE FOR MOVING GOODS WHILE SUBJECTING TO THE ELECTRO-PLATING PROCESS.—Fig. 584, side elevation, with front frame off; Fig. 585, end elevation of that part of front frame where the fly is held; Figs. 586 and 587, the plating vat, with frame moving on inclined plane.

The large wheel A, Fig. 585, is propelled by a weight suspended from the roof by a cord which winds round its barrel, the same as common clockwork. The circumference is studded with small pins which catch the arm B, moving it in a downward direction, and consequently moving the arm C in a forward direction. The latter, being attached to the frame by a small rod R, Figs. 585 and 587, moves it up the inclined plane E, Fig. 587, until the pin fixed in the wheel A passes the end of the arm B. The frame then moving down the incline E, brings B in gear with the next pin, and the same motion again takes place, and so on successively. The speed

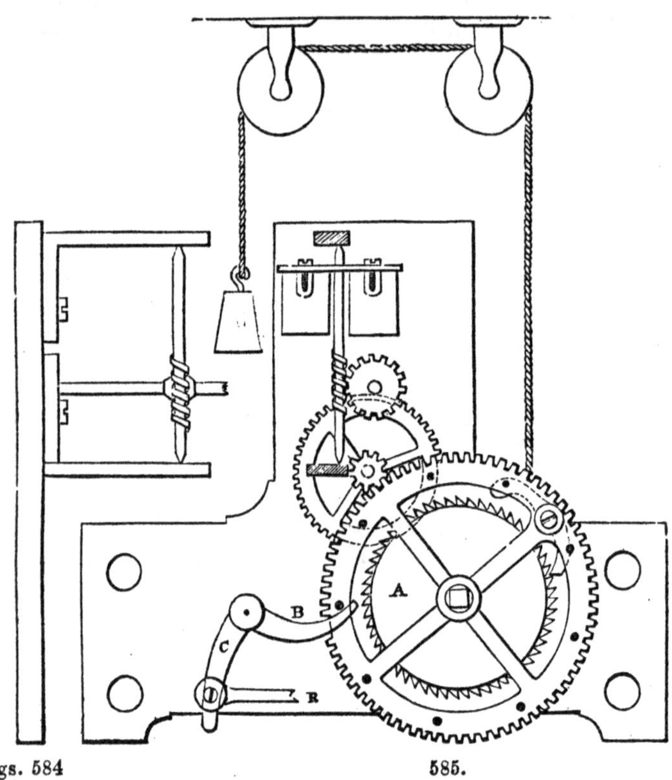

Figs. 584 585.

is regulated by the train moving in an endless screw fitted on the last wheel of the arbor of a fan. The four holes in Fig. 585 are for bolts or pillows for screwing the two frames together. The frame has four pulleys and inclines, the latter adjustable to a greater or less degree by the screw and grove at E.

Deposit dissolving off in Solution.—In depositing any metal, but more particularly such as require solutions having an excess of the solvent, such as of cyanide of potassium in the depositing of gold and silver, care should be taken that nothing stops the current of electricity suddenly, while the article being deposited upon remains in the solution, otherwise the metal deposited will speedily dissolve off. This we have often experienced, and many others have no doubt done the same. Indeed, we have seen a beautiful deposit going on, and left the operation with great hope of excellent results, but on returning shortly after have found the whole dissolved off. And often, when the process was apparently going on well, and the articles had been in the solution the usual time to receive a fair coating of metal, upon taking them out and weighing them there was hardly any perceptible difference from the original weight—in short, there had been no material deposit. These phenomena will be found to occur with the greatest frequency when the solutions and the batteries are in the best condition for working, and when the article upon which the deposit is going on, and the pole or plate of metal forming the positive

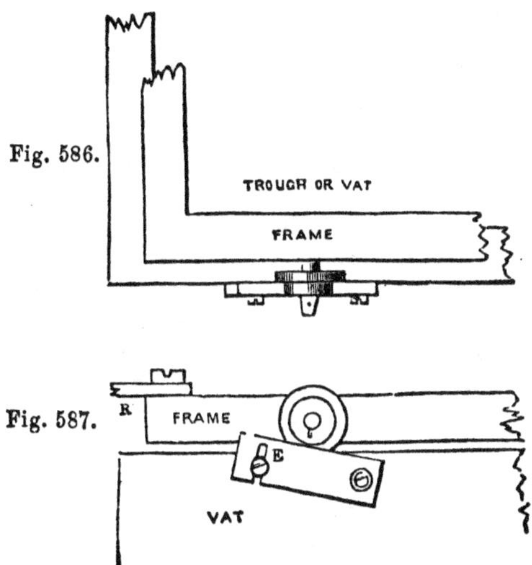

Fig. 586.

Fig. 587.

electrode are at a considerable distance from each other. But before explaining what we consider to be the cause of these annoyances we will refer to another phenomenon connected with them.

Opposite Currents of Electricity from Vats.—If, under the circumstances referred to, and when the deposit has gone on for some time, the wires connecting the battery with the electrodes in the depositing solution be disconnected from the battery, and their two ends be joined together, a current of electricity nearly as strong as that from the battery will pass through the wires, but in the opposite direction from that which was obtained by the battery; and if two pieces of metal were attached to these wires and put into a solution of copper, or any metal, a deposition would occur, the original electrodes now constituting a battery in relation to this second decomposition cell: the current, however, would gradually weaken until it ceased. The cause of all these actions and reactions is this: The article being plated with silver in connection with the battery, exhausts the solution of silver around it, leaving free cyanide of potassium in solution, while around the sheet of silver which is serving as the positive electrode, the solution is on the contrary becoming saturated with silver, so that we have all the conditions necessary to constitute a battery, having silver in two kinds of solution—the one capable of dissolving silver, the other not. In these conditions lies the source of the annoyances described above. From the moment the deposition of metal begins, there also arises an opposite current of electricity, tending to neutralize the effects of the battery, which current goes on increasing in quantity until the two currents neutralize each other, or it may be until the current from the trough overpowers that from the battery. In the latter case, as we have said, there may, at the termination of the ordinary period, be little or no silver deposited on the articles intended to be plated. Motion in the silver or depositing solution will prevent all these annoyances; and this being now generally adopted, these phenomena are not now observed, but the effects take place less or more in every solution.

Test for the Quantity of Free Cyanide of Potassium in Solutions.—It has been already mentioned that the cyanide of silver, as it forms upon the surface of the silver plate, is dissolved by the cyanide of potassium. This renders it necessary to have always in the solution free cyanide of potassium. Were we to use the pure crystalline salt of cyanide of potassium and silver, dissolved in water, without any free cyanide of potassium, we should not obtain a deposit beyond a momentary blush; as the silver plate or electrode would get an instantaneous coating of cyanide of silver, and this not being dissolved the current would stop. The quantity of free cyanide of potassium required in the solution varies according to the amount of silver that is present, and the rapidity of the deposition. If there be too little of it, the deposit will go on slowly; if there be too much, the silver plate will be dissolved in greater proportion than the quantity deposited, and the solution will consequently get stronger. The proportion we have found best is about half by weight of free cyanide of potassium to the quantity of silver in solution; thus, if the solution contains two ounces of silver to the gallon, it should have one ounce of free cyanide of potassium per gallon. This is known by taking some nitrate of silver, dissolving it in distilled water

Fig. 588.

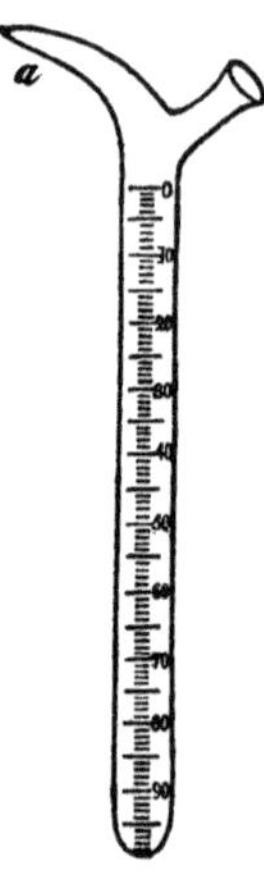

and placing it in a common alkalimeter, graduated into 100 parts, Fig. 588. The proportion of the nitrate of silver in the solution is that every two graduations of the solution should contain 1 grain. A given quantity of the plating solution is now taken—say 1 ounce by measure, and the test solution of nitrate of silver is added to it by degrees, so long as the precipitate formed is redissolved. When this ceases the number of graduations is then noted, and the following equation gives the quantity of free cyanide. Every 175 nitrate of silver are equal to 130 cyanide of potassium in solution. Suppose 20 graduations were taken, equal to 10 grains nitrate of silver, then 175 : 130 : : 10 : 7.4 grains free cyanide of potassium. This, multiplied by 160, the number of fluid ounces per gallon, will make about 2½ ounces. We have taken 2 graduations to one grain of nitrate of silver, that the solution may be considerably dilute and less liable to error. The following table is calculated at a half grain nitrate of silver to the graduation, and will be a guide to the student or workman. The quantity of solution tested is one ounce by measure.

| Number of graduations used. | Free cyanide per gallon. | | |
|---|---|---|---|
| | oz. | dwt. | gr. |
| 1 | 0 | 2 | 13 |
| 2 | 0 | 5 | 3 |
| 3 | 0 | 7 | 16 |
| 4 | 0 | 10 | 6 |
| 5 | 0 | 12 | 19 |
| 6 | 0 | 15 | 9 |
| 7 | 0 | 17 | 22 |
| 8 | 1 | 0 | 13 |
| 9 | 1 | 3 | 1 |
| 10 | 1 | 5 | 12 |
| 11 | 1 | 8 | 5 |
| 12 | 1 | 10 | 19 |
| 13 | 1 | 13 | 8 |
| 14 | 1 | 15 | 22 |
| 15 | 1 | 18 | 11 |
| 16 | 2 | 1 | 2 |
| 17 | 2 | 3 | 14 |
| 18 | 2 | 6 | 2 |
| 19 | 2 | 8 | 11 |
| 20 | 2 | 11 | 0 |

Another method may be adopted. If, for instance, we dissolve a small quantity of sulphate of copper and add to it an excess of ammonia, there is produced a deep blue color. Cyanide of potassium will destroy the blue color, in a fixed chemical proportion. To obtain this proportion, take ten grains of pure cyanide of potassium and dissolve in water; then take a certain quantity, say 100 grains, of sulphate of copper and convert it into ammoniuret, the whole measuring a given quantity, and pour from an alkalimeter this blue liquor into the cyanide of potassium till it ceases to destroy the color, then mark the number of graduations required, and that amount of copper solution will represent 10 grains cyanide of potassium—a quantitative test will thus be got for the full cyanide of potassium in the solution, and should be used as follows: Say that the color of 60 graduations of the blue solution was destroyed by the 10 grains of cyanide of potassium, then to test the quantity of free cyanide of potassium in the plating solution, take 60 graduations of the blue liquor in any convenient vessel, and add to it from an alkalimeter the plating solution till the color of the blue liquor is destroyed, then note the quantity which contains 10 grains free cyanide, from which the quantity in the whole solution may be calculated.

Rate of depositing Silver.—When articles are taken out of the solution they are swilled in water, and then put into boiling water. They are afterwards put into hot sawdust, which dries them perfectly. Their color is chalk-white. They are generally weighed before being scratch-brushed; that is, brushed with the fine wire brushes and stale beer as already described. Although this operation does not displace any of the silver, still, in taking off the chalky appearance, there is a slight loss of weight. The appearance after scratching is that of bright metallic silver. Any thickness of silver may be given to a plate by continuing the operation a proper length of time. One ounce and a quarter to one ounce and a half of silver to the square foot of surface, will give an excellent plate about the thickness of ordinary writing paper.

Bright Deposit.—A little sulphuret of carbon added to the plating solution prevents the chalky appearance, and gives the deposit the appearance of metallic silver; the reaction which takes place in this mixture is not yet understood. The best method of applying the sulphuret of carbon is to put one or two ounces into a large bottle, then fill it with strong silver solution, having an excess of cyanide of potassium, and let it repose for several days, shaking it occasionally. A little of this silver solution is added, as required, to this plating solution, which will give the articles plated the same appearance as if scratched. It is also found that the presence of sulphuret of carbon prevents the solution from going out of order; indeed we have seen a solution that has been constantly working from two to three years, while, generally, they were subject to go out of order for a time, in less than one year—although, after standing a time, they would recover—but these are curious reactions not yet investigated.

Different Metals for Plating.—Silver may be deposited upon any metal, but not upon all with equal facility. Copper, brass, and German silver, are the best metals to plate; iron, zinc, tin, pewter, and Britannia metal, are much more difficult; lead is easier, but it is not a good metal, because of the rapidity with which it tarnishes, and, from its softness, easily yields to the pressure of the burnisher: nevertheless all these metals and alloys may be, and are, plated, but cannot give the satisfaction which brass, copper, or German silver afford. In plating upon alloys having tin in them, such as Britannia metal, they must not be dipped into nitric acid previous to plating; but into a hot and strong solution of caustic potash or soda for five minutes, and put directly into the plating solution (which should have an excess of cyanide of potassium, and the battery be as strong as the liquor will admit of without gas being evolved), until covered, when the silver may be thickened by an ordinary solution and battery.

ELECTRICITY GIVEN OFF FROM SANDY DEPOSITS.—We may mention that when depositing silver upon a large surface, and the solution or battery being in the condition to give the sandy deposit, or rather when the deposit has gone on for a long time and the solution not been agitated, so that it has become very much exhausted of silver round the article, the deposit towards the end of the time has been almost impalpable to the touch, like flour: sometimes the grains were a little coarser. The practice, in such cases, is to lift the articles from the solution, and to place them in boiling water, and after steeping there some time, to take them out, when the heat of the metal soon causes it to dry. Under these circumstances, when the deposit was of the sort stated, we have seen on a large waiter or tray, when the hand was rapidly drawn over the surface, after it was dried in the manner described, the same effect produced as when the hand is drawn over an electrified handkerchief, or sheet of paper, accompanied with a crackling noise and pricking sensation. We have repeatedly observed these phenomena, but never having chanced to be in the dark, no light was visible from the surface rubbed. Although these are the conditions under which the observations were made, the phenomena were not produced every time these conditions were found. It is probably caused by the fact that this kind of deposit, which is of a chalky appearance, is a bad conductor of electricity, and as the boiling water was often very impure, holding salts in solution, the rapid evaporation of the water from the surface of this sort of deposit might leave it excited for a short time, and the hand being drawn across at the time of excitation, the electricity was liberated.

THE OLD METHOD OF PLATING.—Many objections have been urged against the application of electro-deposition to the purposes of plating, as a branch of manufacture, either as a competitor or substitute for the old method, technically called *Sheffield plate*—so called because Sheffield is a principal seat of that manufacture. To enable our readers to form a proper estimate of the objections urged, by enabling them to judge of the relative importance and value of the two processes, we shall add a brief description of the old method.

An ingot of copper being cast, was filed square and smooth, and a piece of silver was placed upon it, the two surfaces being perfectly clean: a little borax having been introduced between the two metals, they were bound together with iron wire, and then heated in a furnace nearly to the melting point; the small quantity of borax increased the fusibility of the two metals at their surface, and thus they were fused together. When fusion was effected the metals were subjected to the dilating process of heavy rollers, the dimensions in length and width being regulated according to the articles to be made. This sheet formed the base or foundation of every article, of whatever shape or form, and however it was to be ornamented when finished.

To produce ornaments, leaf silver was stamped in iron dies representing the ornaments required, which, when removed from the dies, were filled with an alloy of lead and tin. These were then soldered upon the flat or shaped plain surface with soft solder, which melts at a very low temperature: thus were produced the silver edges, or mounts.

The quality of the ornament depended entirely upon the price of the article; but whatever the quality, all ornaments in the old mode of plating were thus made, the only difference being the thickness of the silver leaf used:—Ornamental feet, handles, knobs, etc., were made in the same manner, being struck up in *two* parts, filled with lead and tin, soldered together with soft solder, and afterwards soldered to the main body. Articles (such as table candlesticks) which would be too heavy if filled with lead, were filled with rosin, pitch or any other similar substance, for the purpose of preventing the article being flattened by pressure. Hence it is evident that no *solid* article could be made by the old mode of plating, the only way of producing articles being to work them up by the hammer, or to strike them in dies from a flat surface: and being restricted to the use of soft solder, on account of the plated metal and the shells of silver, forming the edges, not supporting the required heat to melt *silver* solder, it is equally evident that the joinings so constructed would be easily removed either by force or heat.

The nearest approach to *solid* articles made by the old method of plating, were forks and spoons: these were generally made of iron, thin silver being soldered upon the surface, which was afterwards dressed smooth, and polished.

The heat used in this operation was merely that of an ordinary soldering iron; because, were a greater heat applied, the silver would form an alloy with the tin and lead of the solder, and melt: the same heat that cemented the metals in the first instance would be sufficient to disunite them; and thus, when these forks were exposed in hot gravy, the solder was liable to become soft, and the silver covering, yielding readily to the knife, to peel up or become abraided, in consequence of the soft intervening metal.

ADVANTAGES OF ELECTRO-PLATING.—The advantages offered to the plater by the electro-process are many, arising from the fact of the articles being plated *after*, instead of *before*, being manufactured. This at once entirely removed all those restrictions on taste and design, under which the plater was forced, by the nature of his process, to labor.

The following may be considered some of the principal differences existing in the two processes of plating—the old method and the electro-process:

1. The electro-plater is not limited in the use of the metal upon which he plates. There is generally used, as the basis of all electro-plated goods, a hard white metal, which possesses the sound, and approaches very nearly to the color, of silver. Inferior goods are sometimes made in brass.

2. The electro-plater is not restricted to the use of soft solder, which melts at a very low temperature, and forms a very insufficient joint, besides preventing any sound or ring in the article so soldered. Where cheap goods are required, this may be used in *this* process as well as in the old, but is always open to the same objection. All goods of superior quality, made for the electro-process, are soldered with what is termed in the trade *hard silver solder*, composed of 2 parts of silver and 1 part of brass melted together, which is not affected by any ordinary degree of heat, and presents a joint as strong as the metal itself.

The common solder of braziers may also be used with advantage: it is very hard and durable and requires a strong heat to melt it.

3. The electro-plater, in producing ornamental articles, is not obliged to incur the expense of cutting iron dies for every minute portion; being under no restriction, he models his pattern, and by casting and chasing in *solid* metal, produces an exact copy, which is *afterwards* plated or gilt.

Thus any pattern which can be executed in silver may be readily made in plate by this method.

4. The junction of the plating with the metal below, by the electro-process, is perfect, without the presence of any intervening substance: the forks and spoons thus made are not open to the objection of the old process, and are found to answer all the purposes of silver, in sound, appearance, and wear: they are generally tested, previously to polishing, by exposure in a furnace to a red heat.

5. From the facility with which *old* goods may be now restored, these goods bear an intrinsic value; whereas before the introduction of the electro-process, a plated article worn through in any part was valueless.

OBJECTIONS TO ELECTRO-PLATING.—Several objections to the electro-process have been keenly urged; but they may all be reduced to the following:

1st objection: Deposited metal is crystalline, and therefore, though it may impart in appearance a silver coating, it must necessarily be full of minute interstices between the crystals: hence when a metal, such as copper, is plated, it is liable to be acted upon by the atmosphere, or injured by whatever is brought into contact with it

This objection was not without foundation, as all deposited metals are crystalline in texture, but they do not necessarily leave interstices; the objection, however, is almost entirely removed by keeping the articles in motion during the deposition: by motion and proper arrangement of battery we have deposited silver of as high specific gravity as hammered silver, which could not be the case if it were porous.

2d objection: As only pure silver is deposited, it must necessarily be soft, and consequently liable to abrasion, and more rapid wear.

This objection is also partly true. Only pure silver can be deposited; but it is not necessarily soft: the quality of the deposit, in this respect, depends (as already noticed) a great deal upon the nature of the solution and the battery power. Intensity of battery gives hardness to the metal deposited. There is no complaint more common amongst the burnishers of electro-plated articles, than that the metal is hard; and it is far from being an uncommon occurrence, that some goods have to be heated so that they may be more easily burnished or polished. How far this annealing may affect the wear of the goods is not yet ascertained. That the silver is pure we think an advantage—hence the superior color which electro-plated goods possess: besides which, purchasers are not subject to the risk of having a plate much alloyed.

3d objection: The mounts or prominences of articles, which must have the greatest wear, have the least and thinnest deposit.

This objection is entirely without foundation, as the prominences have always the greatest portion of deposit, and the hollow parts the least.

Solid Silver Articles made by the Battery.—Silver may be deposited from its cyanide solution upon wax moulds polished with black lead, almost as easily as copper; but for this purpose it is better to have the solution much stronger in silver than for plating. We have found that 8 ounces of silver to the gallon of solution make a very good strength. Nevertheless, no articles are made in silver by depositing upon wax in this manner. Strong solutions of cyanide of potassium and silver act upon wax, and would soon destroy a mould. The method of making articles in solid silver by the electro-process has been already explained (page 544,) namely, a copper mould is made by the electrotype, and the silver is deposited within this mould to the proper thickness; after which it is kept in a hot solution of crocus and muriatic acid, or boiled in dilute hydrochloric acid, which dissolves the copper without injuring the silver.

The method which we esteem as best for dissolving off the copper is this: an iron solution is first made by dissolving a quantity of copperas in water, placing it on a fire till it begins to boil: a little nitric acid is then added—nitrates of potash and soda will do just as well—the iron, which is thus peroxidized, may be precipitated either by ammonia, or carbonate of soda; the precipitate being washed, muriatic acid is added till the oxide of iron is dissolved. This forms the solution for dissolving the copper. When the solution becomes almost colorless, and has ceased to act on the copper, the addition of a little ammonia will precipitate the iron again; after a little exposure, the copper remains in solution, which is decanted off and preserved for recovering the copper; this is done by neutralizing the ammonia by an acid, and putting in pieces of iron, which deposit the copper in the metallic state. The precipitate of iron is again dissolved in muriatic acid, and employed in dissolving the copper. Thus the iron may be used over and over again with little trouble, and the persalt of iron will be found to dissolve the copper more rapidly than an acid; persulphate of iron must not be used, as it dissolves the silver along with the copper. The silver article is then cleaned in the usual way (page 593), and heated to redness over a clear charcoal fire, which gives it the appearance of dead silver, in which state it may be kept, or, if desired, it may be scratched and burnished.

When ammonia is first added to the above solution of copper and iron, both these metals are precipitated together as a brown precipitate. After a little exposure, the copper dissolves, and the iron is at the same time peroxidized, having been previously reduced to the protoxide by the copper dissolving. When the persulphate of iron is used for dissolving copper, and ammonia is then added to the solution, the same results take place:—the precipitation of both copper and iron. But the compound seems not so stable, the copper passing more quickly into an oxide, which dissolves in the ammonia. However, with free ammonia, either with the chloride or sulphate, the oxidation of the metals is slower than with water alone.

Copper moulds intended for receiving a deposit must be protected on the back, but if the solution is very strong, there is every danger that it will decompose the protecting substance, thus rendering the solution very dirty, and causing a sediment. For the purpose of protecting the mould, various suggestions and experiments have been made; amongst other substances, pitch has been tried: it is easily affected alone, but on boiling a little of it in potash, a heavy and dirty sediment is left, destitute of any adhesive property; on putting a quantity of this sediment into a pot nearly filled with melted pitch, a violent effervescence will take place, setting free a volume of white fumes having a creosotic smell. After all effervescence has ceased, which will not be before a considerable time, and when all the mass seems to have been acted upon, the process of making an excellent protecting coating is completed—a coating which will not yield in the solution, and which is at once both good and cheap, its only fault being its brittleness.

In the manufacture of solid silver articles, the electro-process has not yet been of extensive application: and in making duplicates of rare objects of art, and costly chased or engraved articles in silver, one prevailing objection has been felt, namely, they have no "ring," and seem, when laid suddenly upon a table, to be cracked or unsound, or like so much lead; this disadvantage is no doubt partly owed to the crystalline character of the deposit, and partly to the pure character of the silver, in which state it has not a sound like standard or alloyed silver. That this latter cause is the principal one, appears from the fact that a piece of silver thus deposited is not much improved in sound by being heated and hammered, which would destroy all crystallization.

We may mention that the same objections are applicable to articles made in gold by the electro-deposit; nevertheless, for figures and ornaments, these objections are of little weight. When in Marlborough House a few months back, we were shown a plate of antique pattern in deposited silver, which was all but free from these defects.

Dead Silvering for Medals.—The perfect smoothness which a medal generally possesses on the surface, renders it very difficult to obtain a coating of dead silver upon it, having the beautiful silky lustre which characterizes that kind of work, except by giving it a very thick coating of silver, which takes away the sharpness of the impression. This dead appearance can be easily obtained by putting the medal, previous to silvering, in a solution of copper, and depositing upon it, by means of a weak current, a mere blush of copper, which gives the face of the medal that beautiful crystalline richness that deposited copper is known to give. The medal is then to be washed from the copper solution, and immediately to be put into the silver solution. A very slight coating of silver will suffice to give the dead frosty lustre so much admired, and in general so difficult to obtain.

Oxidizing Silver.—A very beautiful effect is produced upon the surface of silver articles technically termed oxidizing, which gives the surface an appearance of polished steel. This can be easily effected by taking a little chloride of platinum, prepared as described at page 527, heating the solution and applying it to the silver when an oxidized surface is required, and allowing the solution to dry upon the silver. The darkness of the color produced varies according to the strength of the platinum solution, from a light steel gray to nearly black. The effects of this process, when done along with what is termed dead work, is very pretty, and may be easily applied to medals, giving scope for the exercise of taste. Upon this we quote the following:

"The high appreciation in which ornaments in oxidized silver are now held, render a notice of the process followed interesting. There are two distinct shades in use, one produced by chloride, which has a brownish tint, and the other by sulphur, which has a bluish black tint. To produce the former, it is only necessary to wash the article with a solution of sal-ammoniac; a much more beautiful tint may, however, be obtained by employing a solution composed of equal parts of sulphate of copper and sal-ammoniac in vinegar. The fine black tint may be produced by a slightly warm solution of sulphuret of potassium or sodium."—*Chem. Techn. Mittheilungen von Dr. Ellsner.*

Protection of Silver Surface.—All silver or plated articles

are subject to tarnish by exposure to the air, especially in this cli mate, and where coals containing so much sulphur are used; the tarnish being generally a sulphuret of silver. Deposited silver is more easily tarnished than standard silver.

Medals or figures silvered for the sake of their appearance ought to be protected from the air, or they very soon lose their silver color; a medal may be put into a frame air-tight, and a figure should be covered with a glass shade: if the silver has been left dead, any attempt to clean it destroys its appearance. Varnishes have been tried to protect the silver from the atmosphere; but all varnishes, however colorless, detract from the silver lustre, and are not good. For ordinary purposes, medals may be very conveniently protected by laying a piece of common glass over the surface, cut to the exact size, and held close by a piece of paper pasted round the edges of both, and then a stout piece on the back. We have had silver medals, preserved by this means, for more than ten years. Little round medals may be conveniently covered by watch-glasses, fastened on in the same manner.

CLEANING OF SILVER.—A weak solution of cyanide of potassium, used as a wash over tarnished silver, will brighten it. This solution was, and we believe is still, sold in small bottles for this purpose, but it is not good, as it dissolves the silver rapidly, and is such a deadly poison that it must be used with great caution on articles that may be required for domestic purposes.

A variety of cleaning pastes and powders are used for silver or plated goods. Those containing mercury and oxide of lead should be avoided, for although they give a dark color when newly put on, it soon blackens. The best paste we have found is a mixture of fine precipitated chalk, carbonate of magnesia, and oxide of iron. These materials are made into a paste, and rubbed upon the plate with soft leather: for wrought or chased surfaces a hard brush is best. The goods should be finished by polishing with leather and a little of this mixture in a dry state, which will give that fine dark mirror-looking color so much admired. Common coarse whiting and flannel cloths should not be used, as thev wear the silver rapidly.

## CHAPTER XXXI.

### ELECTRO-GILDING.

THE operation of gilding, or covering other metals with a coating of gold, is performed in the same manner as the operation of plating, with the exception of a few practical modifications, which we shall now notice in detail.

PREPARATION OF SOLUTION OF GOLD.—The gold solution for

gilding is prepared by dissolving gold in three parts of muriatic acid and one of nitric acid, which forms the chloride of gold. This is digested with calcined magnesia, and the gold is precipi tated as an oxide. The oxide is boiled in strong nitric acid, which dissolves any magnesia in union with it: the oxide being well washed is dissolved in cyanide of potassium, which gives cyanide of gold and potassium; thus:—

| *Substances used:* | | *Substances produced:* | |
|---|---|---|---|
| Oxide of Gold | $=AuO$ | Cyanide of Gold and Potassium | $=KCy+AuCy$ |
| 2. Cyanide of Potassium | $=2KCy$ | Potash | $=KO$ |

$$Auo+2\ KCy=KO+(KCy+AuCy.)$$

By this method a proportion of potash is formed in the solution, as an impurity; it is not, however, very detrimental to the process. In preparing the oxide of gold there is always a little of the gold lost, to recover which the washings should be kept, evaporated to dryness, and fused.

Another and very simple method is this:—Add a solution of cyanide of potassium to the chloride of gold until all the precipitate is redissolved; but this gives chloride of potassium in the solution, which is not good. In the preparation of the solution by this means there are some interesting reactions. As the chloride of gold has always an excess of acid, the addition of cyanide of potassium causes violent effervescence, and no precipitate of gold takes place until all the free acid is neutralized, which causes a considerable loss to the cyanide of potassium. There is always formed in this deposition a quantity of ammonia and carbonic acid, from the deposition of the cyanate of potash; and if the chloride of gold be recently prepared and hot, there is often formed some aurate of ammonia (fulminate of gold), which precipitates with the cyanide of gold. Were this precipitate to be collected and dried, it would explode when slightly heated. On previously diluting the chloride of gold, or using it cold, this compound is not formed.

After the free acid is neutralized by the potash, further addition of the cyanide of potassium precipitates the gold as cyanide of gold, having a light yellow color; but as this is slightly soluble in ammonia and some of the alkaline salts, it is not advisable to wash the precipitate lest there be a loss of gold. Cyanide of potassium is generally added until the precipitate is redissolved; consequently much impurity is formed in the solution, namely, nitrate and carbonate of potash with chloride of potassium and ammonia. Notwithstanding, this solution works very well for a short time, and it is very good for operations on a small scale.

Battery Process of Preparing Gold Solution.—The best method of preparing the gold solution is that described for silver (p. 589). Say the operator wishes to prepare a gallon of gold solution, he dissolves four ounces of cyanide of potassium in one gallon of water, and heats the solution to 150° Fah.; he now takes a small porous cell and fills it with this cyanide solution, and places it inside the gallon of solution: into this cell is put a small plate of iron or copper, and attached by a wire to the zinc of a battery. A piece of gold is placed into the large solution, facing the plate in the porous cell, and attached to the copper of the battery. The whole is allowed to remain in action until the gold, which is to be taken out from time to time and weighed, has lost the quantity required in solution. By this means a solution of any strength can be made, according to the time allowed. The solution in the porous cell, except the action has continued long, will have no gold, and may be thrown away. Half an hour will suffice for a small quantity of solution—of course any quantity of solution may be made up by the same means. For all the operations of gilding by the cyanide solution, it must be heated to at least 130° Fah. The articles to be gilt are cleaned in the way described for silver, but are not dipped into nitric acid previously to being put in the gold solution. Three or four minutes is sufficient time to gild any small article. After the articles are cleaned and dried they are weighed—and when gilt they are weighed again; thus the quantity of gold deposited is ascertained. Any convenient means may be adopted for heating the solution. The one generally adopted is to put a stoneware pan containing the solution into an iron or tinplate vessel filled with water, which is kept at the boiling point either by being placed upon a hot plate or over gas. The hotter the solution the less battery power is required. Generally three or four pairs of plates are used for gilding, and the solution is kept at 130° to 150° Fah. But one pair will answer if the solution is heated to 200°.

PROCESS OF GILDING.—The process of gilding is generally performed upon silver articles. The method of proceeding is as follows: When the articles are cleaned as described in our chapter on plating, they are weighed, and well scratched with wire brushes, which cleanse away any tarnish from the surface, and prevents the formation of air-bubbles. They are then kept in clean water until it is convenient to immerse them in the gold solution. One immersion is then given, which merely imparts a blush of gold; they are taken out and again brushed; they are then put back into the solution and kept there for three or four minutes, which will be sufficient if the solution and battery are in good condition; but the length of time necessarily depends on these two conditions, which must be studied and regulated by the operator.

Iron, tin, and lead, are very difficult to gild direct; they therefore generally have a thin coating of copper deposited upon them by the cyanide of copper solution and immediately put into the gilding solution.

CONDITIONS REQUIRED IN GILDING.—The gilding solution generally contains from one-half to an ounce of gold in the gallon, but for covering small articles, such as medals, for tinging daguerreotypes, gilding rings, thimbles, etc., a weaker solution will do. The solution should be sufficient in quantity to gild the articles at once, so that it should not have to be done bit by bit; for when there is a part in the solution and a part out, there will generally be a line mark at the point touching the surface of the solution. The rapidity with which metals are acted upon at the surface line of the solution is remarkable. If the positive electrode is not wholly immersed in the solution, it will, in a short time, be cut through at the surface of the water, as if cut by a knife. This is also the case in silver, copper, and other solutions, as before referred to.

MAINTAINING THE GOLD SOLUTION.—As the gold solution evaporates by being hot, distilled water must from time to time be added. The water should always be added when the operation of gilding is over, not when it is about to be commenced, or the solution will not give so satisfactory a result. When the gilding operation is continued successively for several days, the water should be added at night. The average cost of depositing gold is about 4 cts. per pennyweight.

The means of testing the free cyanide of potassium, with nitrate of silver, as described for silver, is not applicable to the gold solution; but it may be tested by the ammoniuret of copper, see p. 603. To obtain a deposit of a good color much depends upon the state of the solution and battery; it is therefore necessary that strict attention be paid to these, and more so as the gold solution is very liable to change if the relative size of the article receiving the deposit is not according to that of the positive plate.

The result of a series of observations and experiments, continued daily throughout a period of nine months, showed that in five instances only the deposit was exactly equal to the quantity dissolved from the positive plate. In many cases the difference did not exceed 3 per cent., though occasionally it rose to 50 per cent. The average difference however was 25 per cent. In some cases double the quantity dissolved was deposited, in others the reverse occurred—both resulting from alterations made in the respective processes; for in these experiments we varied, as far as practicable, the state of the solution and the relative sizes of the negative and positive electrodes.

The most simple method of keeping a constant register of the state of the solution is to weigh the gold electrode before putting it into the solution; and, when taking it out, to compare the loss with the amount deposited: a little allowance, however, must be made for small portions of metal dissolved in the solution, from the articles that are gilt, which, when gilding is performed daily, is considerable in a year. A constant control can thus be exercised over the solution, to which there will have to be added from time to time a little cyanide of potassium, a simple test of requirement being that the gold pole should always come out clean—for if it has a film or crust it is a certain indication that the solution is deficient of cyanide of potassium. Care must be taken to distinguish this crust, which is occasionally dark-green or black, from a black appearance which the gold pole will take when very small in comparison to the article being gilt, and which is caused by the tendency to evolve gas. In this case an addition of cyanide of potassium would increase the evil. The black appearance from the tendency to the escape of gas has a slimy appearance. This generally takes place when the solution is nearly exhausted of gold, of which fact this appearance, taken conjointly with the relative sizes of the electrodes, are a sure guide.

To Regulate the Color of the Gilding.—The gold upon the gilt article, on coming out of the solution should be of a dark-yellow color, approaching to brown, but this when scratched will yield a beautifully-rich deep gold. If the color is blackish it ought not to be finished, for it will never either brush or burnish a good color. If the battery is too strong, and gas is given off from the article, the color will be black; if the solution is too cold, or the battery rather weak, the gold will be light-colored; so that every variety of shade may be imparted. A very rich dead gild may be made by adding ammoniuret of gold to the solution just as the articles are being put in, or what is better, add some sulphuret of carbon in the same way as for silver solutions, which affects the color and appearance of the gold in the same way as it does the silver.

Coloring of Gilding.—A defective colored gilding may be improved by the same method as that adopted in the old process to color gilding or gold, namely, by the help of the following mixture:

3 parts Nitrate of Potash
1½ Alum
1½ Sulphate of Zinc
1½ Common Salt.

These ingredients are put into a small quantity of water, to form a sort of paste, which is put upon the articles to be colored: they are then placed upon an iron plate over a clear fire, so that they will attain nearly to a black heat, when they are suddenly plunged into cold water: this gives them a beautiful high color. Different hues may be had by a variation in the mixture.

To Dissolve Gold from Gilt Articles.—Before regilding articles which are partly covered with gold, or when the gilding is imperfect, and the articles require regilding, the gold should be removed from them by putting them into strong nitric acid; and when the articles have been placed in the acid, by adding some common salt, not in solution, but in crystals. By this method gold may be dissolved from any metal, even from iron, without injuring it in the least. After coming out of the acid, the articles must be polished. The best method, however, is to brush off the gold as described for silver (page 597), which gives the polish at the same time.

To Recover the Gold.—When the gold is dissolved off by the acid after it is saturated, or when it ceases to dissolve the gold rapidly, the acid is diluted with several times its bulk of water, and then soda or potash added till the greater portion of the acid is neutralized. A solution of sulphate of iron (copperas) is then added, so long as a precipitate is formed; when this settles down it is carefully collected upon a paper filter, washed and dried, and then fused in a crucible with a little borax and common salt, when the gold is found as a button at the bottom of the crucible.

When the gold is brushed off, the brushings are burned at a red heat, and the residue fused with carbonate of soda and a little borax: in this case, the gold will not be pure and will have to be refined.

Objections to Electro-gilding.—Objections have also been made to the application of electro-gilding to the arts, of the same kind as those urged against electro-plating; but the now almost universal adoption of this process by gilders, because of the perfection to which the articles are brought, forms the best answer we can give to such objections. However, let us take a hasty glance at the old process and its consequences, that we may be enabled to judge of the comparative merits of both methods.

Before the introduction of electro-deposition, the only method of gilding was by forming an amalgam of gold and mercury, which, at the consistence of a thin paste, was brushed upon the articles over a strong heat; the mercury being gradually dissipated, the gold remained fixed upon the articles. This process is most pernicious, and destructive to human life; the mercury, volatilized by the heat, insinuates itself into the bodies of the workmen, notwithstanding the greatest care; and those who are so fortunate as to escape for a time absolute disease, are constantly liable to salivation from its effects. Paralysis is common among them, and the average of their lives is very short; it has been estimated as not exceeding 35 years. It is difficult to believe that men could be found to engage in such a business, reckless of the consequences so fearfully exhibited before them; and it would naturally be thought they would hail with pleasure the introduction of any process which would be put a stop to such a dreadful sacrifice of human life. But it is very difficult to overcome interest and prejudice, even when the object to be gained is of such vast importance.

Effects of Cyanogen on Health.—The effects produced upon the health of those who work constantly over cyanide solutions are not yet fully tested, by which we could form a comparison with the old process; for every new trade, or operation, gives rise to a new disease, or some new forms of an old disease. Having ourselves inhaled much of the fumes of that "*ominous*" gas given off from the cyanide of potassium solution, we are not prepared to stand its advocate, but would rather warn all employed at the business, or who may in any degree have to do with these solutions, to be very careful not to use too much freedom. The hands of those engaged in gilding or plating are subjected to ulceration, particularly if they have been immersed in the solution. The ulcers are not only annoying but painful; and, on their first appearance, if care is not properly taken to wash them in strong cyanide of potassium, and then in acid water, the operator will, in a short time, have to take a few days rest. We have repeatedly seen, by the aid of a magnifying glass, gold and silver reduced in these ulcerations. We have also known of eruptions breaking out over the bodies of workmen after inhaling those deleterious fumes when they were very bad, as when solutions were precipitated by acids or being evaporated to dryness in a close apartment for the recovery of the metal. Repeatedly have we seen the legs of workmen thus afflicted, and always after they have been exposed to extra fumes.

The following statement of the general effects of electro-plating and gilding on the health of those engaged in them, as experienced by ourselves and others, may not be uninteresting to our readers: but it is necessary to premise that the apartments in which we were employed were improperly ventilated.

The gas has a heavy sickening smell, and gives to the mouth a saline taste, and scarcity of saliva; the saliva secreted is frothy. The nose becomes dry and itchy, and small pimples are found within the nostrils, which are very painful (we have felt these effects in the nose from the hydrogen of the batteries, where there were no cyanide solutions). Then follows a general languor of body; disinclination to take food, and a want of relish. After being in this state for some time, there follows a benumbing sensation in the head, with pains, *not* acute, shooting along the brow; the head feels as a heavy mass, without any individuality in its operations. Then there is bleeding at the nose in the mornings when newly out of bed; after that comes giddiness; objects are seen flitting before the eyes, and momentary feelings as of the earth lifting up, and then leaving the feet, so as to cause a stagger. This is accompanied with feelings of terror, gloomy apprehension, and irritability of temper. Then follows a rushing of blood to the head; the rush is felt behind the ears with a kind of hissing noise, causing severe pain and blindness: this passes off in a few seconds, leaving a giddiness which lasts for several minutes. In our own case the rushing of blood was without pain, but attended with instant blindness, and then followed with giddiness. For months afterwards a dimness remained as if a mist intervened between us and the objects looked at: it was always worse towards evening, when we grew very languid and inclined to sleep. We rose comparatively well in the morning: yet were restless, our stomach was acid, visage pale, features sharp, eyes sunk in the head, and round them dark in color: these effects were slowly developed. Our experience was nearly three years.

We have been thus particular in detailing these effects, as a warning to all employed in the process; but we have no doubt that in lofty rooms, airy and well ventilated, these effects would not be felt. Employers would do well to look to this matter; and amateurs, who only use a small solution in a tumbler, should not, as the custom sometimes is, keep it in their bed-rooms; the practice is decidedly dangerous.

Practical Suggestions in Gilding.—According to the amount of gold deposited, so will be its durability: a few grains will serve to give a gold *color* to a very large surface, but it will not last: this proves, however, that the process may be used for the most inferior quality of gilding. Gold thinly laid upon silver will be of a light color, because of the property of gold to transmit light. The solution for gilding silver should be made very hot, but for copper it should be at its minimum heat. A mere blush may be sufficient for articles not subjected to wear; but on watch-cases, pencil-cases, chains, and the like, a good coating should be given. An ordinary sized watch-case should have from 20 grains to a pennyweight; a mere coloring will be sufficient for the inside, but the outside should have as much as possible. A watch-case thus gilt, for ordinary wear, will last five or six years without becoming bare. We have known some to be in use full six years without losing their covering. Small silver chains, such as those sold at eight shillings, should have 12 grains; pencil-cases, of ordinary size should have from 3 to 5 grains; a thimble from 1 to 2 grains. These suggestions will serve as a guide to amateur gilders, many of whom, having imparted only a color to their pencil-cases, feel chagrin and disappointment upon seeing them speedily become bare; hence arises much of the obloquy thrown upon the process.

# CHAPTER XXXII.

## RESULTS OF EXPERIMENTS ON THE DEPOSITION ON OTHER METALS AS COATINGS.

Coating with Platinum.—This metal has never yet been successfully deposited as a protecting coating to other metals. A solution may be made by dissolving it in a mixture of nitric and muriatic acids, the same as is employed in dissolving gold; but heat must be applied. The solution is then evaporated to dryness, and to the remaining mass is added a solution of cyanide of potassium; next, it must be slightly heated for a short time, and then filtered This solution, evaporated, yields beautiful crystals of cyanide of platinum and potassium; but it is unnecessary to crystallize the salt. A very weak battery power is required to deposit the metal: the solution should be heated to 100°. Great care must be taken to obtain a fine metallic deposit: indeed the operator may not succeed once in twenty times in getting more than a mere coloring of metal over the surface, and that not very adhesive. The causes of the difficulty are probably these: the platinum used as an electrode is not acted upon; the quantity of salt in solution is very little; it requires a particular battery strength to gives a good deposit, and the slightest strength beyond this gives a black deposit; so that, were the proper relations obtained, whenever there is any deposit, the relations of battery and solution are changed, and the black pulverulent deposit follows.

We have occasionally succeeded in obtaining a bright metallic deposit of platinum, possessing the qualities of adhesion and durability: some of the articles thus covered presented no signs of change after many years: but we have never been so fortunate as to get a platinum deposit that could protect any metal from the action of acids, or other fluids by which the metal could be affected. We have covered iron, such as the end of a glass-blower's blow-pipe, so that it could be made red-hot without the iron rusting, but rather taking the characteristic appearance of platinum: but even that did not protect the iron from rusting when it was put a short time into water, or kept exposed to moist air. We have seen again and again recommendations of certain solutions of platinum for the purpose of obtaining a reguline metal, and no doubt it has been obtained, but, as stated above, we believe more incidentally than at will. The protoxalate of platinum has been strongly recommended for covering copper and brass with platinum.*

COATING WITH PALLADIUM.—Palladium is a metal very easily deposited. The solution is prepared by dissolving the metal in nitro-muriatic acid, and evaporating the solution nearly to dryness; then adding cyanide of potassium till the whole is dissolved: the solution is then filtered and ready for use. The cyanide of potassium holds a large quantity of this metal in solution, and the electrode is acted upon while the deposit is proceeding. Articles covered with this metal assume the appearance of the metal; but so far as we are aware, it has not yet been applied to any practical purpose. It requires rather a thick deposit to protect metals from the action of acids, which is, probably, the only use it can be applied to.

COATING WITH NICKEL.—Nickel is very easily deposited; and may be prepared for this purpose by dissolving it in nitric acid, then adding cyanide of potassium to precipitate the metal; after which the precipitate is washed and dissolved by the addition of more cyanide of potassium. Or the nitrate solution may be precipitated by carbonate of potash; this should be well washed, and then dissolved in cyanide of potassium; a proportion of carbonate of potash will be in the solution, which we have not found to be detrimental. This latter method of preparing the nickel plating solution is simple, and, therefore, has our recommendation. The

* Polytechni. Constah. 1855.

metal is very easily deposited; it yields a color approaching to silver, which is not liable to tarnish on exposure to the air. A coating of this metal would be very useful for covering common work such as gasaliers, and other gas-fittings, and even common plate. The great difficulty experienced is to obtain a positive electrode: the metal is very difficult to fuse, and so brittle that we have never been able to obtain either a plate or a sheet of it. Could this difficulty be easily overcome, the application of nickel to the coating of other metals would be extensive, and the property of not being liable to tarnish would make it eminently useful for all general purposes. We coated articles with nickel in 1845, which were exposed to the air for many years without tarnish, and when last seen by the author exhibited no change.

ANTIMONY, ARSENIC, TIN, IRON, LEAD, BISMUTH, AND CADMIUM.—We have deposited these metals from their solutions in cyanide of potassium; but not for any useful application.

IRON.—Iron may be very easily deposited from its sulphate: dissolve a little crystalline sulphate of iron in water, and add a few drops of sulphuric acid to the solution: one pair of Smee's battery may be used to deposit the iron upon copper or brass. The metal in this pure state has a very bright and beautiful silver color.

LEAD.—Lead may be deposited from a solution of an acid salt, such as the acetate, but requires some management or strength of battery: it may also be deposited from its solution in potash or soda.

ALUMINIUM AND SILICIUM.—Since the publication of the former edition of this work, new methods have been discovered for obtaining the base or metal of alumina and silica, or clay and sand, in the metallic state possessing extraordinary properties. One of the methods successfully adopted, is by fusing in a small crucible some chloride or fluoride of aluminium, and when in fusion, inserting two steel poles in connection with a battery which reduces the salt, giving small globules of the metal aluminium.

Attempts have also been made to deposit the metals from their cyanous solution as coating upon other metals in the usual way. We have not ourselves tried any experiments upon these metals, but we take the following results of experiments from Mr. G. Gore of Birmingham, who seems to have given the subject a good deal of attention:

"It has long been known to chemists that all kinds of clay, stone, and sand, of which the earth is composed, consist of metals combined with oxygen, carbonic acid, sulphuric acid, and other non-metallic elements, forming therewith oxides, carbonates, sulphates, etc.; thus clay is an oxide of aluminium, sand an oxide of silicium, limestone a carbonate of calcium; but the separation of the metallic bases from the non-metallic elements with which they are combined has been a matter of so great difficulty, that few chemists have put themselves to the trouble of accomplishing it, and those who have done so have made use of the most powerful means and reducing agents, such as large voltaic batteries, potassium, etc., and have then obtained them in a state of alloy or combination with mercury. Sir Humphrey Davy, the discoverer of most of these bases, in his experiments on the decomposition of the alkalies and earths, used a powerful battery, consisting of 500 pairs of plates, and then succeeded in obtaining them combined with mercury, from which they were afterwards separated. Wöhler and Berzelius, in their discoveries of the means of separating the metals aluminium and silicium from their respective compounds, clay and sand, used a high temperature and potassium, and then succeeded in obtaining them in the condition of dull metallic powders, nearly infusible.

"By a means recently discovered, and described in the March number of the *Philosophical Magazine* for this year, I have succeeded in depositing the metals aluminium from clay, and silicium from sandstone, each in a perfect metallic condition, by dissolving pipe clay, common red sand, pounded stone, etc., in various chemical liquids, and passing currents of electricity from ordinary small voltaic batteries through the solutions.

"My attention has since been directed to produce simple processes, whereby any person not possessing a knowledge of chemistry may readily coat articles with those metals, and thus cause the discovery to be immediately applied to human benefit in the arts and manufactures, and the following are the results of my experiments:

"To coat articles of copper, brass, or German silver, with aluminium, take equal measures of sulphuric acid and water, or take one measure each of sulphuric and hydrochloric acids and two measures of water; add to the water a small quantity of pipe-clay, in the proportion of 5 or 10 grs. by weight to every ounce by measure of water (or ½ oz. to the pint): rub the clay with the water until the two are perfectly mixed, then add the acid to the clay solution, and boil the mixture in a covered glass vessel one hour. Allow the liquid to settle, take the clear, supernatant solution, while hot, and immerse in it an earthen porous cell, containing a mixture of one measure of sulphuric acid and ten measures of water, together with a rod or plate of amalgamated zinc; take a small Smee's battery, of three or four pairs of plates, connected together intensity fashion, and connect its positive pole by a wire, with a piece of zinc in the porous cell. Having perfectly cleaned the surface of the article to be coated, connect it by a wire with the negative pole of the battery, and immerse it in the hot clay solution; immediately abundance of gas will be evolved from the whole of the immersed surface of the article, and in a few minutes, if the size of the article is adapted to the quantity of the current of electricity passing through it, a fine white deposit of aluminium will appear all over the surface. It may then be taken out, washed quickly in clean water, and wiped dry, and polished; but if a thicker coating is required, it must be taken out when the deposit becomes dull in appearance, washed, dried, polished, and re-immersed; and this must be repeated at intervals, as often as it becomes dull, until the required thickness is obtained. With small articles it is not absolutely necessary, either in this or the following process, that a separate battery be employed, as the article to be coated may be connected by a wire with a piece of zinc in the porous cell, and immersed in the outer liquid, when it will receive a deposit, but more slowly than when a battery is employed.

"To coat articles with silicium, take the following proportions: three-quarters of an ounce, by measure, of hydrofluoric acid, ¼ oz. of hydrochloric acid, and 40 or 50 grs. either of precipitated silica, or of fine white sand (the former dissolves most freely), and boil the whole together for a few minutes, until no more silica is dissolved. Use this solution exactly in the same manner as the clay solution, and a fine white deposit of metallic silicium will be obtained, provided that the size of the article is adapted to the quantity of the electric current: common red sand, or indeed any kind of silicious stone, finely powdered, may be used in place of the white sand, and with equal success, if it be previously boiled in hydrochloric acid, to remove the red oxide of iron or other impurities.

"Both in depositing aluminium and silicium, it is necessary to well saturate the acid with the solid ingredients by boiling, otherwise very little deposit of metal will be obtained.

"Among the many experiments I have made upon this subject, the following are a few of the most interesting:—

"*Experiment* 1.—Boiled some pipe-clay in caustic potash and water, poured the clear part of the solution into a glass vessel, and immersed in it a small earthen porous cell, containing dilute sulphuric acid and a piece of amalgamated zinc; immersed a similar piece of bright sheet copper in the alkaline liquid, and connected it with the negative pole of a small Smee's battery of three pairs of plates, connected the zinc plate with the positive pole, and let the whole stand undisturbed all night; on examining it next morning I found the piece of copper coated with a white silver-like deposit of metallic aluminium.

"*Experiment* 2.—Obtained from a railway cutting in the town a small piece of the sand rock upon which Birmingham is built, boiled it in hydrochloric acid, to remove the red oxide of iron, washed it clean with water, and dissolving it by boiling it in a mixture of hydrofluoric acid, nitric acid, and water; immersed in this solution, a porous cell with dilute acid and zinc, as before; connected a piece of brass with the zinc by a wire, and suspended it in the outer liquid, which was kept hot by a small spirit lamp beneath; after allowing the action to proceed several hours, I found the piece of brass beautifully coated with white metallic silicium.

"*Experiment* 3.—Took one part, by weight, of the same sandstone, after being purified by the hydrochloric acid, and 2½ parts of carbonate of potash, fused them together in a crucible until all evolution of gas ceased, and a perfect glass was formed; poured out the melted glass, and when cold, dissolved it in water, and used this solution in the same manner as the former ones, allowing the action to proceed about 12 hours, when a good white deposit of metallic silicium was obtained.

"*Experiment* 4.—Took some stones with which the streets of Birmingham are macadamised, pounded them fine in a mortar, boiled the powder in hydrochloric acid, to purify it from iron, washed it well in water, and dissolved it by boiling an excess of it in a mixture of ¾ oz., by measure, of hydrofluoric acid, ½ oz. of water, and ½ oz. each of nitric and hydrochloric acids, until no more would dissolve; used the clear portion of this solution in the same manner as the former liquids, and readily coated in it a piece of brass with a beautifully white deposit either of aluminium or silicium. From these, and many other experiments which I have tried, it is quite clear that common metal articles may be readily coated with white metals, possessing similar characters to silver, from solutions of the most common and abundant materials, and thus bring within the purchase of the poorer classes articles of taste and cleanliness which are at present only to be obtained by the comparatively wealthy."

TIN.—Tin is easily deposited from a solution of protochloride of tin. If the two poles or electrodes be kept about two inches apart, a most beautiful phenomenon may be observed. The decomposition of the solution is so rapid that it shoots out from the negative electrode like tentacula, or feelers, towards the positive, which it reaches in a few seconds. The space between the poles seems like a mass of crystallized threads, and the electric current passes through them without effecting further decomposition. So tender are these metallic threads that when lifted out of the solution they fall upon the plate like cobweb. Seen through a glass they exhibit a beautiful crystalline structure. If a circular electrode of tin is used, and a small wire put in the centre of the chloride solution, the thread-like crystals will shoot out all round, and give quite a metallic confervæ. Tin may also be deposited from its solution in caustic potash or soda.

ANTIMONY.—In the deposition of antimony Mr. Gore has observed a curious and interesting phenomenon, that the metal during its deposition, and after some has been deposited, explodes occasionally, the particles being thrown about by the shock.

DEPOSITION OF ALLOYS.—Many attempts have been made to deposit alloys of metals from their solutions. That two or more metals can be deposited from a solution we have seen sufficient evidence; but the means to regulate the proportions of each, and to make such a process practical, have yet to be discovered. It is hardly possible to get a mixed solution of any two metals that are exactly equally decomposable; or, in other words, that the metals under the circumstances in which they are placed are exactly of equal conducting power. Hence the electric current will always travel through the one that offers the least resistance, and there will be none of the other metallic solution decomposed, or metal deposited, until the quantity of electricity is greater than the best conducting metal in the solution will allow to pass; then the other metal will be deposited in proportion to the extra electrical power that passes. As, for example, take a mixture of cyanide of gold, silver, and copper, in cyanide of potassium. The silver in this state is so much superior in its conducting power to the other salts, that all the silver may be deposited from the solution by a weak battery without any of the other metals. If the solution be afterwards heated, and the battery power kept so that no gas is allowed to escape from the articles, the gold may be deposited without any copper; but if the gas is allowed to flow from the article receiving the deposit, the copper will be deposited, and often more abundantly than the gold, as the escape of gas is not consistent with a reguline deposit of gold. We have thus deposited an alloy of gold and copper; we have also deposited gold and silver, but the alloy was very inferior and irregular. Alloys can be obtained from silver and palladium, from cyanide solutions, from zinc and copper, from a solution of their sulphates; but in no instance have we found good alloys, or alloys that could pass as such in name or appearance. We have seen articles, such as iron, covered with copper and zinc in this manner, or in alternate layers, and the articles having the coating heated in charcoal, by which means a brass of fair appearance was obtained, but the process is attended with practical difficulty, and the product cannot be called deposited brass.

Several patents have been taken out for the deposition of alloys of various sorts. The following, by Morris and Johnson, embraces a wide range, and being well described we will copy the specification:

"This invention consists in the employment of solutions composed of cyanide of potassium and carbonate of ammonia, to which are added cyanides, carbonates, and other compounds of metals, in proportions according to the amount of deposit required to be made.

"In order that the invention may be fully understood and carried into effect, the patentees proceed to describe the means pursued by them as follows: These improvements consist in the employment of solutions composed of carbonate of ammonia (the carbonate of ammonia of commerce, or the sesqui-carbonate of ammonia of chemists), and cyanide of potassium, to which are added carbonates, cyanides, or other compounds of metals, in various proportions. For the well-known alloy, brass, carbonate of ammonia and cyanide of potassium are used in the following proportions, namely—To each or every gallon of water are added 1 lb. of carbonate of ammonia, 1 lb. of cyanide of potassium, 2 ozs. of cyanide of copper, and 1 oz. of cyanide of zinc. These proportions may be varied to a considerable extent. Or the patentees take the before-named solution of carbonate of ammonia and cyanide of potassium, in the proportion of 1 lb. of each to one gallon of water; and they take a large sheet of brass of the desired quality, and make it the anode or positive electrode, in the aforesaid solution, of a powerful galvanic battery or magneto-electric machine, and a small piece of metal, and make it the cathode or negative electrode, from which hydrogen must be freely evolved. This operation is continued till the solution has taken up a sufficient quantity of the brass to produce a reguline deposit. The solution may be used cold; but it is desirable, in many cases, to heat it (according to the nature of the article or articles to be deposited upon) up to 212° Fahr. For wrought or fancy work about 150° Fahr. will give excellent results. The galvanic battery, or magneto-electric machine, must be capable of evolving hydrogen freely from the cathode or negative electrode, or article attached thereto. It is preferred to have a large anode or positive electrode, as this favors the evolution of hydrogen. The article or articles treated as before described will immediately become coated with brass. By continuing the process any desired thickness may be obtained. Should the copper have a tendency to come down in a greater proportion than is desired, which may be known by the deposit assuming too red an appearance, it is corrected by the addition of carbonate of ammonia, or by a reduction of temperature, when the solution is heated. Should the zinc have a tendency to come down in too great a proportion, which may be seen by the deposit being too pale in its appearance, this is corrected by the addition of cyanide of potassium or by an increase of temperature.

"The alloy, German silver, is deposited by means of a solution consisting of carbonate of ammonia and cyanide of potassium (in the proportions previously given for the brass), and cyanides or other compounds of nickel, copper, and zinc, in the requisite proportions to constitute German silver. It is however preferred to make the solution by means of the galvanic battery or magneto-electric machine, as above described for brass. Should the copper of the German silver come down in too great a proportion, this is corrected by adding carbonate of ammonia, which brings down the zinc more freely; and should it be necessary to bring down the copper in greater quantity, cyanide of potassium is added—such treatment being similar to that of the brass before described.

"The solutions for the alloys of gold, silver, and other alloys of metals, are made in the same manner as above stated, by employing anodes of the alloy or alloys to be deposited; or by adding to the solutions the carbonates, cyanides or other compounds, in the proportions forming the various alloys—always using in depositing an anode of the required alloy. These solutions are subject to the same treatment and control as those of the brass and German silver before described.

"The patentees claim the combination of the carbonate of ammonia, before named or other carbonates of ammonia and cyanide of potassium, as the ingredients for their solutions for depositing alloys of metals."

The expense of depositing common metals will remain a barrier to the use of electro-metallurgy in making such alloys as brass, a consideration which some patentees do not seem to consider. We have the following given as methods for mixing up solutions for depositing brass, which will prove our position:

"As an illustration of this invention we take the patentee's method of depositing a coating of brass by galvanic agency, in which he employs the following:—1. A solution of the double chloride of zinc and ammonia.—2. A solution of the double chloride of zinc and potassium.—3. A solution of the double chloride of zinc and sodium.—4. A solution of the double acetate of zinc and ammonia.—5. A solution of the double acetate of zinc and potassium.—6. A solution of the acetate of zinc and soda.—7. A saturated solution of carbonate of zinc, and carbonate of ammonia. —8. A solution of the double tartrate of zinc, and of potash, soda, or ammonia. (To one thousand parts of the solution of tartrate of zinc, indicating three degrees on the salinometer, thirty parts of hydrochlorate of ammonia, and eighty parts of hydrochloric acid must be added.)—9. A solution of citrate of zinc rendered soluble by an excess of citric acid.—10. A solution of tartrate of zinc in potash or soda. With each of the above solutions an analogous solution of copper must be mixed in the proportion suitable for obtaining the required depth of color."

Besides the ordinary electro-metallurgical operations, the public are from time to time told through the press that the process has been applied to the extraction of metals from their ores, but on examination the statement is invariably found to be incorrect, the metal being in all cases separated from the ore by means of an acid or acids, and the electro-metallurgical operations not applied till after this separation takes place; so that its application is altogether apart from the extraction of the metal from the ore. Such an application for common metals is *commercially* absurd; and nothing can exhibit the want of practical application so much as some of the patents taken out for this object. The greater number of these patents are intended for copper ores, upon which we will offer a few remarks. It will be seen from the principles of deposition, that, allowing the copper was all in solution to be deposited by a battery, the cheapest form known will give the loss of one ton of zinc and sulphuric acid to get one ton of copper, which would be upwards of £20 for the materials destroyed, while a ton of copper may be smelted by the ordinary process for half that sum. We give the following extract of a patent as an illustration, not because it is worse than others, but being more definite in its methods and battery than most of these patents, and the patentee an excellent electrician.

"Mr. Andrew Crosse, of Broomfield, the electrician, has just specified his patent for improvements in the extraction of metals from their ores. The apparatus employed for this purpose consists of a wooden or earthenware vessel capable of holding from 250 to 300 quarts, at a short distance above the bottom of which is a movable platinum frame covered with a netting of platinum wire, the meshes being about 1 inch each way. This frame is connected to the positive pole of a Daniell's battery by a platinum wire, covered with a non-conducting material throughout those parts of it exposed to the liquid in the vessel; the negative pole of the battery being connected to a copper wire, from which is suspended by three smaller wires, in the interior of the vessel, a bowl of wood lined with sheet copper and covered with a copper wire netting. The battery in connection with the apparatus should consist of 20 pairs of plates, each in a gallon glass vessel, filled with a saturated solution of sulphate of copper, to which has been added from 1-20th to 1-10th part of sulphuric acid.

The mode of operating is as follows: The vessel is partially filled with water acidulated with sulphuric acid—230 quarts of water and 5 quarts of sulphuric acid being a convenient quantity. About 15 lbs. of the copper ore, previously calcined and reduced to powder, is then stirred into the liquid in the vessel and allowed to subside, after which the platinum frame is lowered on to the surface of the ore, and the copper-lined bowl suspended in its place, when the electric current immediately begins to act; but it is preferred to allow the ore to remain four or five days in the acidulated water before applying the electric current. The liquid during the process should be kept heated even as high as the boiling point, by which the separation of the copper and its deposition in the bowl will be facilitated. The time occupied in effecting this is generally three or four days, when the whole of the copper is removed. The acid liquid and sediment, which will contain any other metals that may have been present, are run out through a plug-hole in the bottom of the vessel. The sediment should be tested to ascertain if it still contains any proportion of copper; and if so, it can be mixed with fresh calcined ore and again operated on. The liquid does not require any fresh quantity of acid to be added to it during the process, and afterwards it may again be similarly used.

Here we have 20 pairs of plates recommended to be used in the battery, making a destruction of 20 tons of zinc and acid for one ton of copper, and taking four days to deposit. Twenty-one tons of copper per week would be but a small quantity of copper made, compared with smelting; and at the ordinary per centage of ore to get this, there will have to be operated upon 300 tons of ore, requiring acres of tanks, heated according to specification, independent of the furnace for calcining. Having got the ore calcined and free of sulphur, it would be preferable to fuse it with carbonaceous matters and get the copper direct. Notwithstanding the commercial absurdity of all these applications and patents, still there are several ingenious adaptations worthy of the attention of the electro-metallurgist as a study in his profession.

DEPOSITION OF BRONZE.—The following solutions of different metals are given by Brunel, Bisson, and Gaugain, as being capable of giving a deposit of bronze:

| | | |
|---|---|---|
| 50 | parts | Carbonate of Potash. |
| 2 | " | Chloride of Copper |
| 4 | " | Sulphate of Zinc. |
| 25 | " | Nitrate of Ammonia. |

A bronze plate is used as the positive electrode. The deposit given by this solution has been seen by Becquerel, who mentions that it bears comparison with any ordinary bronze in appearance.* A solution of the above materials in water strikes the ear as somewhat hypothetical: that a mixed solution of copper and zinc will give, under certain conditions, a compound deposit we know, and also that, with a quantity of other salts present, will give peculiar tints of color, a circumstance which may be obtained without a compound deposit. But the difficulty to be overcome is to proportion the deposit of different metals, so that we may make up a solution and battery that will deposit either Mantz's yellow metal, Stirling's yellow metal, gun metal, or common brass, at pleasure; and that we may be able to produce compounds that are constant and unvarying: so that, for example, we could deposit silver or gold of the standard quality, all which, notwithstanding the many statements that have been made in print, have yet to be discovered.

---

We have thus given a brief review of the practical operations of Electro-Metallurgy for the guidance of the student, who as he proceeds, will find that the difficulties which at first beset his path will gradually disappear: easier modifications of processes will suggest themselves, as all operators cannot with equal facility follow the same directions. New facts will reveal themselves to his inquiries; a wide field of interesting and profitable research will open up before his mind; and the steady and persevering experimenter and observer will not fail to reap an abundant harvest of honor and gratification, in being an instrument in promoting the knowledge of the working of the laws of Nature.

---

* Progress of General Science, vol. ii.

## CHAPTER XXXIII.

### THEORETICAL OBSERVATIONS.

WE have described at considerable length the practical details connected with the art of electro-metallurgy, without pausing to inquire into the philosophy of the action of the electric currents by which the effects are produced. It will be unnecessary to enter into a long discussion of the numerous theories that have been advanced from time to time to explain the action that takes place in a battery or decomposing cell, while the current is passing through the solution—a brief reference to the more commonly received opinions being sufficient for the present purpose.

ACTION OF SULPHATE OF COPPER ON IRON.—In order to convey our ideas accurately, let us suppose that the solution undergoing decomposition is sulphate of copper. This salt is composed of sulphuric acid and copper, which may be represented as $SO^4 + Cu$: these are held together according to the law of chemical affinity; but if iron is put into the solution, the combination of the acid and copper will be dissolved by the attraction of the acid to the iron, for which it has a stronger affinity than for the copper. Hence iron, put into sulphate of copper, decomposes it thus:

$$Cu, SO^4 + Fe = Fe, SO^4 + Cu.$$

Were we to put a piece of copper into a solution of sulphate of copper, there would be no action, the forces being equal; but if by any means we were to communicate to this piece of copper a higher attractive force for the $SO^4$ than that of the copper which is already in union with it, we should cause the acid to leave the copper it was originally combined with, and to combine with the new piece of copper. Bearing these general principles in view, we shall proceed to state the different opinions of authors on this subject.

FARADAY'S THEORY OF ELECTROLYSIS.—Professor Faraday says—"Passing to the consideration of electro-chemical decomposition, it appears to me that the effect is produced by an *internal corpuscular* action, excited according to the direction of the electric current, and that it is due to a force either *superadded to*, or *giving direction to*, the *ordinary* chemical affinity of the bodies present. The body under decomposition (say sulphate of copper), may be considered as a mass of acting particles, all those which are included in the course of the electric current contributing to the final effect; and it is because the ordinary chemical affinity is relieved, weakened, or partly neutralized by the influence of the electric current in one direction parallel to the course of the latter, and strengthened or added to in the opposite direction, that the combining particles have a tendency to pass in opposite courses.

"In this view the effect is considered as *essentially dependent* upon the *mutual chemical affinity* of the particles of opposite kinds. Particles *aa* could not be transferred or travel from one pole N, towards the other pole P, unless they found particles of the opposite kind, *bb*, ready to pass in the contrary direction; for it is by virtue of their increased affinity for those particles, combined with their diminished affinity for such as are behind them in their course, that they are urged forward.

Fig. 589.

"I conceive the effects to arise from forces which are *internal*, relative to the matter under decomposition, and not *external*, as they might be considered, if directly dependent upon the poles. I suppose that the effects are due to a modification by the electric current of the chemical affinity of the particles, through or by which that current is passing, giving them the power of acting more forcibly in one direction than in another, and consequently making them travel by a series of successive decompositions, in opposite directions, and finally causing their expulsion or exclusion at the boundaries of the body under decomposition, in the direction of the current, *and that* in larger or smaller quantities, according as the current is more or less powerful."*

In the above figure, the particles *aa* may be termed copper Cu, and the particles *bb*, sulphuric acid $SO^4$, which will enable us to follow the comparison of the different views.

GRAHAM'S THEORY OF ELECTROLYSIS.—Professor Graham supposes that the compound particles, such as sulphate of copper, possess polarity, so that the particles in the battery or decomposition cell will stand in relation to each other in a polar chain, as in Fig. 590.

Fig. 590.

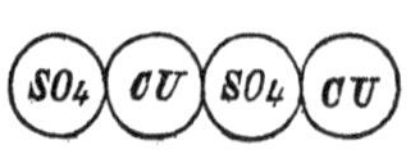

$SO_4$ CU $SO_4$ CU

He then represents electrotyping by the porous cell system, as follows:

"The liquids on either side of the porous division may also be different, provided they have both a polar molecule. Thus, in Fig. 591 the polar chain is composed of molecules of hydrochloric acid, extending from the zinc to the porous division at *a*, and of molecules of chloride of copper from *a*, to the copper plate. When the Cl of molecule 1 unites with zinc, the H of that molecule unites with the Cl of molecule 2 (as indicated by the connecting bracket below); the H of molecule 2 with the Cl of molecule 3; the Cu of molecule 3 with the Cl of molecule 4; and the Cu of this molecule being the last in the chain, is deposited upon the copper plate. Dilute sulphuric acid in contact with an amalgamated zinc plate, and the same acid fluid saturated with sulphate of copper in contact with the copper plate, are a combination of fluids of most frequent application."† According to this theory

* Faraday's Experimental Researches, vol. i. paragraphs 518, 519, 524
† Graham's Elements of Chemistry, 2d edition, 1859.

all the particles between the zinc and copper during the action of the batteries will be performing a whirling motion; for, when the Cl of molecule 1 is liberated, the H of 1 will combine the Cl of 2, which compound molecule must whirl round to be in its proper polar position, which will necessitate that interchange distinctly referred to by Professor Faraday—a mutual transfer of the elements; the Cl will pass towards the zinc plate, and the H and Cu towards the copper plate.

Fig. 591.

Zinc. Copper.

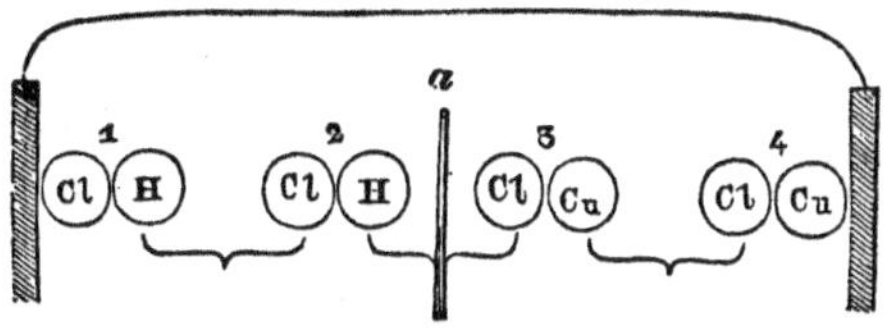

DANIELL'S AND MILLER'S VIEWS.—Theories varying little from these were held by the late Professor Daniell, till, by a series of interesting experiments, in company with Professor Miller, he found that there is no mutual transfer of the elements; that the negative element, or that represented above as Cl or $SO^4$, is transferred from the copper to the zinc, or in a decomposition cell from the negative electrode to the positive electrode: but the positive element—that represented by H or Cu—is not transferred; therefore, the theories of Professors Faraday and Graham are opposed to a fundamental truth experimentally proved. Professors Daniell and Miller conclude their paper, read before the Royal Society, by the following observations:

"These facts are, we believe, irreconcilable with any of the molecular hypotheses which have been hitherto imagined to account for the phenomena of electrolysis, nor have we any more satisfactory at present to substitute for them; we shall therefore prefer leaving them to the elucidation of further investigations to adding one more to the already too numerous list of hasty generalizations."*

In this paper, the authors state that they found certain positive elements transferred in small proportions; thus, potassium from sulphate of potash in $\frac{1}{3}$ of an equivalent; barium, from nitrate of barytes, $\frac{1}{6}$ equivalent; and magnesia, from sulphate of magnesia, $\frac{1}{12}$ equivalent. This was a difficulty for forming any theory, but we have shown that this difficulty does not exist.

In all cases where two liquids are separated by a porous diaphragm, there is a mutual transfer of the liquids in distinct ratios, according to time, either by what is called endosmosis, or by a diffusion; and the rate of transfer is materially affected by a galvanic current passing through them. From observations and operations made on a large scale, and from experiments on various kinds of

* Philosophical Transactions, Part I. for 1844.

solutions, we believe that the fractional transfers of Professors Daniell and Miller are the results of endosmosis or diffusion, and not of electrolytic transfer. According to recent experiments by Professor Graham, diffusion takes place in definite proportions. We believe that no transfer of any base or positive element takes place by electrolysis.

Proposed Theory.—Having carefully considered the various phenomena attending electrolysis, in the decomposition of metallic salts, we think that the electricity is conducted through the solution by the base, or positive element, in the electrolyte, which it does as if it was a solid chain of particles—or wire. We have already said, that if to a solution of sulphate of copper we put a piece of iron, the acid in union with the copper will leave it and combine with the iron. If a piece of copper be put into the same solution, no change will take place; but if we by any means give to this copper an increased tendency to unite with the acid, it will attract the acid from the copper in solution by virtue of this increased attraction. Suppose two wires coming from a battery are

Fig. 592.

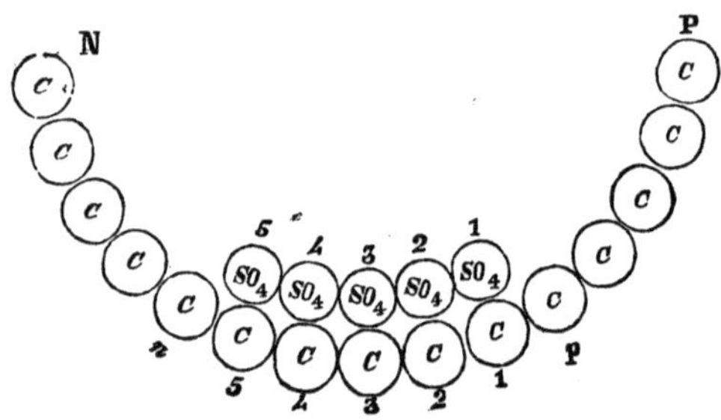


placed in a solution of sulphate of copper, thus, (Fig. 592): the double row representing the compound atoms of sulphate of copper forming the electrolyte: C C the copper or positive element, and $SO^4$ the sulphuric acid or negative element of the solution. The two single rows C C, etc., at each end of the double row, represent the wire or solid conductors of the electricity, from the battery to the decomposition cell: the last particle of the single rows *p n* nearest the double row may be viewed as the electrodes. The sulphuric acid $SO^4$, and the copper C, in solution, are held together by their affinity for each other.

Now let it be supposed that an equivalent of electricity leaves the positive terminal of the battery P, and passes along the solid particles of the conductor, that particle upon which the electricity is, must be for the time in a higher state of excitement than the other particles. When the electric current comes to the last particle of the solid chain *p*, which is in contact with the electrolyte, its increased excitement causes it to attract and combine with the acid particle $SO^4$ nearest it; the electricity being dynamic, passes to the first basic particle C1, giving it an exalted excitement, which causes it to unite with the acid particle $SO^4$2, the electric force passing to C2, which becomes excited in turn, and takes the particles $SO^4$3; and so on through the chain till the last particle C5, which, having no further acid to combine with, gives its electricity to the solid conductor, or electrode *n*, and passes along to the battery, the particle C5 being thus left adhering to the solid chain of particles, or electrode.

By this we observe that every equivalent of decomposition will carry an equivalent of acid to the positive electrode, without taking the metallic element to the other or opposite electrode. This is exactly the fact of the case, the result that takes place in all solutions undergoing decomposition by the current, and also in the battery between the zinc and copper. In these explanations we have spoken of electricity as a material substance, passing along the line, being more easily conceived than the theory of vibrations, etc.; but the effects are the same, and we have seen no phenomena in electric decomposition which are inconsistent with the views here given.

# INDEX.

# CATALOGUE
OF
# PRACTICAL AND SCIENTIFIC BOOKS,
PUBLISHED BY
## HENRY CAREY BAIRD,
INDUSTRIAL PUBLISHER,
NO. 406 WALNUT STREET,
PHILADELPHIA.

---

☞ Any of the Books comprised in this Catalogue will be sent by mail, free of postage, at the publication price.

☞ MY NEW AND ENLARGED CATALOGUE, 95 pages 8vo., with full descriptions of Books, will be sent, free of postage, to any one who will favor me with his address.

---

**ARMENGAUD, AMOUROUX, AND JOHNSON.—THE PRACTICAL DRAUGHTSMAN'S BOOK OF INDUSTRIAL DESIGN, AND MACHINIST'S AND ENGINEER'S DRAWING COMPANION:** Forming a complete course of Mechanical Engineering and Architectural Drawing. From the French of M. Armengaud the elder, Prof. of Design in the Conservatoire of Arts and Industry, Paris, and MM. Armengaud the younger and Amouroux, Civil Engineers. Rewritten and arranged, with additional matter and plates, selections from and examples of the most useful and generally employed mechanism of the day. By WILLIAM JOHNSON, Assoc. Inst. C. E., Editor of "The Practical Mechanic's Journal." Illustrated by 50 folio steel plates and 50 wood-cuts. A new edition, 4to. . $10 00

**ARLOT.—A COMPLETE GUIDE FOR COACH PAINTERS.** Translated from the French of M. ARLOT, Coach Painter; late Master Painter for eleven years with M. Ehrler, Coach Manufacturer, Paris. With important American additions . . $1 25

**ARROWSMITH.—PAPER-HANGER'S COMPANION:** A Treatise in which the Practical Operations of the Trade are Systematically laid down: with Copious Directions Preparatory to Papering; Preventives against the Effect of Damp on Walls; the Various Cements and Pastes adapted to the Several Purposes of the Trade; Observations and Directions for the Panelling and Ornamenting of Rooms, &c. By JAMES ARROWSMITH. 12mo., cloth . . . . . . $1 25

**BAIRD.—THE AMERICAN COTTON SPINNER, AND MANAGER'S AND CARDER'S GUIDE:**

A Practical Treatise on Cotton Spinning; giving the Dimensions and Speed of Machinery, Draught and Twist Calculations, etc.; with notices of recent Improvements: together with Rules and Examples for making changes in the sizes and numbers of Roving and Yarn. Compiled from the papers of the late ROBERT H. BAIRD. 12mo. . . . . $1 50

**BAKER.—LONG-SPAN RAILWAY BRIDGES:**

Comprising Investigations of the Comparative Theoretical and Practical Advantages of the various Adopted or Proposed Type Systems of Construction; with numerous Formulæ and Tables. By B. Baker. 12mo. . . . . . . $2 00

**BAKEWELL.—A MANUAL OF ELECTRICITY—PRACTICAL AND THEORETICAL:**

By F. C. BAKEWELL, Inventor of the Copying Telegraph. Second Edition. Revised and enlarged. Illustrated by numerous engravings. 12mo. Cloth . . . .

**BEANS.—A TREATISE ON RAILROAD CURVES AND THE LOCATION OF RAILROADS:**

By E. W. BEANS, C. E. 12mo. . . . $2 00

**BLENKARN.—PRACTICAL SPECIFICATIONS OF WORKS EXECUTED IN ARCHITECTURE, CIVIL AND MECHANICAL ENGINEERING, AND IN ROAD MAKING AND SEWERING:**

To which are added a series of practically useful Agreements and Reports. By JOHN BLENKARN. Illustrated by fifteen large folding plates. 8vo. . . . . . . $9 00

**BLINN.—A PRACTICAL WORKSHOP COMPANION FOR TIN, SHEET-IRON, AND COPPER-PLATE WORKERS:**

Containing Rules for Describing various kinds of Patterns used by Tin, Sheet-iron, and Copper-plate Workers; Practical Geometry; Mensuration of Surfaces and Solids; Tables of the Weight of Metals, Lead Pipe, etc.; Tables of Areas and Circumferences of Circles; Japans, Varnishes, Lackers, Cements, Compositions, etc. etc. By LEROY J. BLINN, Master Mechanic. With over One Hundred Illustrations. 12mo. $2 50

**BOOTH.—MARBLE WORKER'S MANUAL:**
Containing Practical Information respecting Marbles in general, their Cutting, Working, and Polishing; Veneering of Marble; Mosaics; Composition and Use of Artificial Marble, Stuccos, Cements, Receipts, Secrets, etc. etc. Translated from the French by M. L. BOOTH. With an Appendix concerning American Marbles. 12mo., cloth . . $1 50

**BOOTH AND MORFIT.—THE ENCYCLOPEDIA OF CHEMISTRY, PRACTICAL AND THEORETICAL:**
Embracing its application to the Arts, Metallurgy, Mineralogy, Geology, Medicine, and Pharmacy. By JAMES C. BOOTH, Melter and Refiner in the United States Mint, Professor of Applied Chemistry in the Franklin Institute, etc., assisted by CAMPBELL MORFIT, author of "Chemical Manipulations," etc. Seventh edition. Complete in one volume, royal 8vo., 978 pages, with numerous wood-cuts and other illustrations. $5 00

**BOWDITCH.—ANALYSIS, TECHNICAL VALUATION, PURIFICATION, AND USE OF COAL GAS:**
By Rev. W. R. BOWDITCH. Illustrated with wood engravings. 8vo. . . . . . . . . . $6 50

**BOX.—PRACTICAL HYDRAULICS:**
A Series of Rules and Tables for the use of Engineers, etc. By THOMAS BOX. 12mo. . . . . . . $2 50

**BUCKMASTER.—THE ELEMENTS OF MECHANICAL PHYSICS:**
By J. C. BUCKMASTER, late Student in the Government School of Mines; Certified Teacher of Science by the Department of Science and Art; Examiner in Chemistry and Physics in the Royal College of Preceptors; and late Lecturer in Chemistry and Physics of the Royal Polytechnic Institute. Illustrated with numerous engravings. In one vol. 12mo. . $1 50

**BULLOCK.—THE AMERICAN COTTAGE BUILDER:**
A Series of Designs, Plans, and Specifications, from $200 to to $20,000 for Homes for the People; together with Warming, Ventilation, Drainage, Painting, and Landscape Gardening. By JOHN BULLOCK, Architect, Civil Engineer, Mechanician, and Editor of "The Rudiments of Architecture and Building," etc. Illustrated by 75 engravings. In one vol. 8vo. . . . . . . . . . . $3 50

**BULLOCK.—THE RUDIMENTS OF ARCHITECTURE AND BUILDING:**

For the use of Architects, Builders, Draughtsmen, Machinists, Engineers, and Mechanics. Edited by JOHN BULLOCK, author of "The American Cottage Builder." Illustrated by 250 engravings. In one volume 8vo. . . . $3 50

**BURGH.—PRACTICAL ILLUSTRATIONS OF LAND AND MARINE ENGINES:**

Showing in detail the Modern Improvements of High and Low Pressure, Surface Condensation, and Super-heating, together with Land and Marine Boilers. By N. P. BURGH, Engineer. Illustrated by twenty plates, double elephant folio, with text. $21 00

**BURGH.—PRACTICAL RULES FOR THE PROPORTIONS OF MODERN ENGINES AND BOILERS FOR LAND AND MARINE PURPOSES.**

By N. P. BURGH, Engineer. 12mo. . . . $2 00

**BURGH.—THE SLIDE-VALVE PRACTICALLY CONSIDERED:**

By N. P. BURGH, author of "A Treatise on Sugar Machinery," "Practical Illustrations of Land and Marine Engines," "A Pocket-Book of Practical Rules for Designing Land and Marine Engines, Boilers," etc. etc. etc. Completely illustrated. 12mo. . . . . . . . . . . . . $2 00

**BYRN.—THE COMPLETE PRACTICAL BREWER:**

Or, Plain, Accurate, and Thorough Instructions in the Art of Brewing Beer, Ale, Porter, including the Process of making Bavarian Beer, all the Small Beers, such as Root-beer, Ginger-pop, Sarsaparilla-beer, Mead, Spruce beer, etc. etc. Adapted to the use of Public Brewers and Private Families. By M. LA FAYETTE BYRN, M. D. With illustrations. 12mo. $1 25

**BYRN.—THE COMPLETE PRACTICAL DISTILLER:**

Comprising the most perfect and exact Theoretical and Practical Description of the Art of Distillation and Rectification; including all of the most recent improvements in distilling apparatus; instructions for preparing spirits from the numerous vegetables, fruits, etc.; directions for the distillation and preparation of all kinds of brandies and other spirits, spirituous and other compounds, etc. etc.; all of which is so simplified that it is adapted not only to the use of extensive distillers, but for every farmer, or others who may wish to engage in the art of distilling. By M. LA FAYETTE BYRN, M. D. With numerous engravings. In one volume, 12mo. $1 50

**BYRNE.—POCKET BOOK FOR RAILROAD AND CIVIL ENGINEERS:**
Containing New, Exact, and Concise Methods for Laying out Railroad Curves, Switches, Frog Angles and Crossings; the Staking out of work; Levelling; the Calculation of Cuttings; Embankments; Earth-work, etc. By OLIVER BYRNE. Illustrated, 18mo., full bound . . . . . . $1 75

**BYRNE.—THE HANDBOOK FOR THE ARTISAN, MECHANIC, AND ENGINEER:**
By OLIVER BYRNE. Illustrated by 185 Wood Engravings. 8vo. $5 00

**BYRNE.—THE ESSENTIAL ELEMENTS OF PRACTICAL MECHANICS:**
For Engineering Students, based on the Principle of Work. By OLIVER BYRNE. Illustrated by Numerous Wood Engravings, 12mo. . . . . . . . . . . $3 63

**BYRNE.—THE PRACTICAL METAL-WORKER'S ASSISTANT:**
Comprising Metallurgic Chemistry; the Arts of Working all Metals and Alloys; Forging of Iron and Steel; Hardening and Tempering; Melting and Mixing; Casting and Founding; Works in Sheet Metal; the Processes Dependent on the Ductility of the Metals; Soldering; and the most Improved Processes and Tools employed by Metal-Workers. With the Application of the Art of Electro-Metallurgy to Manufacturing Processes; collected from Original Sources, and from the Works of Holtzapffel, Bergeron, Leupold, Plumier, Napier, and others. By OLIVER BYRNE. A New, Revised, and improved Edition, with Additions by John Scoffern, M. B , William Clay, Wm. Fairbairn, F. R. S., and James Napier. With Five Hundred and Ninety-two Engravings; Illustrating every Branch of the Subject. In one volume, 8vo. 652 pages . $7 00

**BYRNE.—THE PRACTICAL MODEL CALCULATOR:**
For the Engineer, Mechanic, Manufacturer of Engine Work, Naval Architect, Miner, and Millwright. By OLIVER BYRNE. 1 volume, 8vo., nearly 600 pages . . . . . $4 50

**BEMROSE.—MANUAL OF WOOD CARVING:** With Practical Illustrations for Learners of the Art, and Original and Selected designs. By WILLIAM BEMROSE, Jr. With an Introduction by LLEWELLYN JEWITT, F. S. A., etc. With 128 Illustrations. 4to., cloth . . . . . . . . . . . . . $3 00

**BAIRD.—PROTECTION OF HOME LABOR AND HOME PRODUCTIONS NECESSARY TO THE PROSPERITY OF THE AMERICAN FARMER:**

By HENRY CAREY BAIRD. 8vo., paper . . . . . 10

**BAIRD.—THE RIGHTS OF AMERICAN PRODUCERS, AND THE WRONGS OF BRITISH FREE TRADE REVENUE REFORM.**

By HENRY CAREY BAIRD. (1870) . . . . . 5

**BAIRD.—SOME OF THE FALLACIES OF BRITISH-FREE-TRADE REVENUE-REFORM.**

Two Letters to Prof. A. L. Perry, of Williams College, Mass. By HENRY CAREY BAIRD. (1871.) Paper . . . . . 5

**BAIRD.—STANDARD WAGES COMPUTING TABLES:**

An Improvement in all former Methods of Computation, so arranged that wages for days, hours, or fractions of hours, at a specified rate per day or hour, may be ascertained at a glance. By T. SPANGLER BAIRD. Oblong folio . . . . . . $5 00

**BAUERMAN.—TREATISE ON THE METALLURGY OF IRON.**

Illustrated. 12mo. . . . . . . . . . $2 50

**BICKNELL'S VILLAGE BUILDER.**

55 large plates. 4to. . . . . . . . . . $10 00

**BISHOP.—A HISTORY OF AMERICAN MANUFACTURES:**

From 1608 to 1866; exhibiting the Origin and Growth of the Principal Mechanic Arts and Manufactures, from the Earliest Colonial Period to the Present Time; By J. LEANDER BISHOP, M. D., EDWARD YOUNG, and EDWIN T. FREEDLEY. Three vols. 8vo., $10 00

**BOX.—A PRACTICAL TREATISE ON HEAT AS APPLIED TO THE USEFUL ARTS:**

For the use of Engineers, Architects, etc. By THOMAS BOX, author of "Practical Hydraulics." Illustrated by 14 plates, containing 114 figures. 12mo. . . . . . . . . $4 25

**CABINET MAKER'S ALBUM OF FURNITURE:**

Comprising a Collection of Designs for the Newest and Most Elegant Styles of Furniture. Illustrated by Forty-eight Large and Beautifully Engraved Plates. In one volume, oblong $5 00

**CHAPMAN.—A TREATISE ON ROPE-MAKING:**

As practised in private and public Rope-yards, with a Description of the Manufacture, Rules, Tables of Weights, etc., adapted to the Trade; Shipping, Mining, Railways, Builders, etc. By ROBERT CHAPMAN. 24mo. . . . . . . . . . . $1 50

**CRAIK.—THE PRACTICAL AMERICAN MILLWRIGHT AND MILLER.**

Comprising the Elementary Principles of Mechanics, Mechanism, and Motive Power, Hydraulics and Hydraulic Motors, Mill-dams, Saw Mills, Grist Mills, the Oat Meal Mill, the Barley Mill, Wool Carding, and Cloth Fulling and Dressing, Wind Mills, Steam Power, &c. By DAVID CRAIK, Millwright. Illustrated by numerous wood engravings, and five folding plates. 1 vol. 8vo. . . . . . $5 00

**CAMPIN.—A PRACTICAL TREATISE ON MECHANICAL ENGINEERING:**

Comprising Metallurgy, Moulding, Casting, Forging, Tools, Workshop Machinery, Mechanical Manipulation, Manufacture of Steam-engines, etc. etc. With an Appendix on the Analysis of Iron and Iron Ores. By FRANCIS CAMPIN, C. E. To which are added, Observations on the Construction of Steam Boilers, and Remarks upon Furnaces used for Smoke Prevention; with a Chapter on Explosions. By R. Armstrong, C. E., and John Bourne. Rules for Calculating the Change Wheels for Screws on a Turning Lathe, and for a Wheel-cutting Machine. By J. LA NICCA. Management of Steel, including Forging, Hardening, Tempering, Annealing, Shrinking, and Expansion. And the Case-hardening of Iron. By G. EDE. 8vo. Illustrated with 29 plates and 100 wood engravings. $6 00

**CAMPIN.—THE PRACTICE OF HAND-TURNING IN WOOD, IVORY, SHELL, ETC.:**

With Instructions for Turning such works in Metal as may be required in the Practice of Turning Wood, Ivory, etc. Also an Appendix on Ornamental Turning. By FRANCIS CAMPIN, with Numerous Illustrations, 12mo., cloth . . $3 00

**CAPRON DE DOLE.—DUSSAUCE.—BLUES AND CARMINES OF INDIGO.**

A Practical Treatise on the Fabrication of every Commercial Product derived from Indigo. By FELICIEN CAPRON DE DOLE. Translated, with important additions, by Professor H. DUSSAUCE. 12mo.

**CAREY.—THE WORKS OF HENRY C. CAREY:**

CONTRACTION OR EXPANSION? REPUDIATION OR RESUMPTION? Letters to Hon. Hugh McCulloch. 8vo. 38

FINANCIAL CRISES, their Causes and Effects. 8vo. paper 25

HARMONY OF INTERESTS; Agricultural, Manufacturing, and Commercial. 8vo., paper . . . . . . $1 00
Do. do. cloth . . . $1 50

LETTERS TO THE PRESIDENT OF THE UNITED STATES. Paper . . . . . . . . . . . . $1 00

MANUAL OF SOCIAL SCIENCE. Condensed from Carey's "Principles of Social Science." By KATE MCKEAN. 1 vol. 12mo. . . . . . . . . . . . $2 25

MISCELLANEOUS WORKS: comprising "Harmony of Interests," "Money," "Letters to the President," "French and American Tariffs," "Financial Crises," "The Way to Outdo England without Fighting Her," "Resources of the Union," "The Public Debt," "Contraction or Expansion," "Review of the Decade 1857—'67," "Reconstruction," etc. etc. 1 vol. 8vo., cloth . . . . . . . . . . . $4 50

MONEY: A LECTURE before the N. Y. Geographical and Statistical Society. 8vo., paper . . . . . . 25

PAST, PRESENT, AND FUTURE. 8vo. . . . . $2 50

PRINCIPLES OF SOCIAL SCIENCE. 3 volumes 8vo., cloth $10 00

REVIEW OF THE DECADE 1857—'67. 8vo., paper 50

RECONSTRUCTION: INDUSTRIAL, FINANCIAL, AND POLITICAL. Letters to the Hon. Henry Wilson, U. S. S. 8vo. paper . . . . . . . . . 50

THE PUBLIC DEBT, LOCAL AND NATIONAL. How to provide for its discharge while lessening the burden of Taxation. Letter to David A. Wells, Esq., U. S. Revenue Commission. 8vo., paper . . . . . . . . . 25

THE RESOURCES OF THE UNION. A Lecture read, Dec. 1865, before the American Geographical and Statistical Society, N. Y., and before the American Association for the Advancement of Social Science, Boston . . . . 50

THE SLAVE TRADE, DOMESTIC AND FOREIGN; Why it Exists, and How it may be Extinguished. 12mo., cloth $1 50

LETTERS ON INTERNATIONAL COPYRIGHT. (1867.) Paper . . . . . . . . . . . 50

REVIEW OF THE FARMERS' QUESTION. (1870.) Paper 25

RESUMPTION! HOW IT MAY PROFITABLY BE BROUGHT AROUT. (1869.) 8vo., paper . . . . . 50

REVIEW OF THE REPORT OF HON. D. A. WELLS, Special Commissioner of the Revenue. (1869.) 8vo., paper 50

SHALL WE HAVE PEACE? Peace Financial and Peace Political. Letters to the President Elect. (1868.) 8vo., paper 50

THE FINANCE MINISTER AND THE CURRENCY, AND THE PUBLIC DEBT. (1868.) 8vo., paper . . 50

THE WAY TO OUTDO ENGLAND WITHOUT FIGHTING HER. Letters to Hon. Schuyler Colfax. (1865.) 8vo., paper $1 00

WEALTH! OF WHAT DOES IT CONSIST? (1870.) Paper 25

**CAMUS.—A TREATISE ON THE TEETH OF WHEELS:**

Demonstrating the best forms which can be given to them for the purposes of Machinery, such as Mill-work and Clock-work. Translated from the French of M. CAMUS. By JOHN I. HAWKINS. Illustrated by 40 plates. 8vo. . . . . . . $3 00

**COXE.—MINING LEGISLATION.**

A paper read before the Am. Social Science Association. By ECKLEY B. COXE. Paper . . . . . . . 20

**COLBURN.—THE GAS-WORKS OF LONDON:**

Comprising a sketch of the Gas-works of the city, Process of Manufacture, Quantity Produced, Cost, Profit, etc. By ZERAH COLBURN. 8vo., cloth . . . . . . . . 75

**COLBURN.—THE LOCOMOTIVE ENGINE:**

Including a Description of its Structure, Rules for Estimating its Capabilities, and Practical Observations on its Construction and Management. By ZERAH COLBURN. Illustrated. A new edition. 12mo. . . . . . . . . $1 25

**COLBURN AND MAW.—THE WATER-WORKS OF LONDON:**

Together with a Series of Articles on various other Waterworks. By ZERAH COLBURN and W. MAW. Reprinted from "Engineering." In one volume, 8vo. . . $4 00

**DAGUERREOTYPIST AND PHOTOGRAPHER'S COMPANION:**

12mo., cloth . . . . . . . . . . $1 25

**DIRCKS.—PERPETUAL MOTION:**

Or Search for Self-Motive Power during the 17th, 18th, and 19th centuries. Illustrated from various authentic sources in Papers, Essays, Letters, Paragraphs, and numerous Patent Specifications, with an Introductory Essay by HENRY DIRCKS, C. E. Illustrated by numerous engravings of machines. 12mo., cloth . . . . . . . . . . $3 50

**DIXON.—THE PRACTICAL MILLWRIGHT'S AND ENGINEER'S GUIDE:**

Or Tables for Finding the Diameter and Power of Cogwheels; Diameter, Weight, and Power of Shafts; Diameter and Strength of Bolts, etc. etc. By THOMAS DIXON. 12mo., cloth. $1 50

**DUNCAN.—PRACTICAL SURVEYOR'S GUIDE:**

Containing the necessary information to make any person, of common capacity, a finished land surveyor without the aid of a teacher. By ANDREW DUNCAN. Illustrated. 12mo., cloth. $1 25

**DUSSAUCE.—A NEW AND COMPLETE TREATISE ON THE ARTS OF TANNING, CURRYING, AND LEATHER DRESSING:**

Comprising all the Discoveries and Improvements made in France, Great Britain, and the United States. Edited from Notes and Documents of Messrs. Sallerou, Grouvelle, Duval, Dessables, Labarraque, Payen, René, De Fontenelle, Malapeyre, etc. etc. By Prof. H. DUSSAUCE, Chemist. Illustrated by 212 wood engravings. 8vo. . . . . . $10 00

**DUSSAUCE.—A GENERAL TREATISE ON THE MANUFACTURE OF SOAP, THEORETICAL AND PRACTICAL:**

Comprising the Chemistry of the Art, a Description of all the Raw Materials and their Uses. Directions for the Establishment of a Soap Factory, with the necessary Apparatus, Instructions in the Manufacture of every variety of Soap, the Assay and Determination of the Value of Alkalies, Fatty Substances, Soaps, etc. etc. By PROFESSOR H. DUSSAUCE. With an Appendix, containing Extracts from the Reports of the International Jury on Soaps, as exhibited in the Paris Universal Exposition, 1867, numerous Tables, etc. etc. Illustrated by engravings. In one volume 8vo. of over 800 pages . . . . . . . . . . $10 00

**DUSSAUCE.—PRACTICAL TREATISE ON THE FABRICATION OF MATCHES, GUN COTTON, AND FULMINATING POWDERS.**

By Professor H. DUSSAUCE. 12mo. . . . $3 00

**DUSSAUCE.—A PRACTICAL GUIDE FOR THE PERFUMER:** Being a New Treatise on Perfumery the most favorable to the Beauty without being injurious to the Health, comprising a Description of the substances used in Perfumery, the Formulæ of more than one thousand Preparations, such as Cosmetics, Perfumed Oils, Tooth Powders, Waters, Extracts, Tinctures, Infusions, Vinaigres, Essential Oils, Pastels, Creams, Soaps, and many new Hygienic Products not hitherto described. Edited from Notes and Documents of Messrs. Debay, Lunel, etc. With additions by Professor H. DUSSAUCE, Chemist. 12mo. $3 00

**DUSSAUCE.—A GENERAL TREATISE ON THE MANUFACTURE OF VINEGAR, THEORETICAL AND PRACTICAL.** Comprising the various methods, by the slow and the quick processes, with Alcohol, Wine, Grain, Cider, and Molasses, as well as the Fabrication of Wood Vinegar, etc. By Prof. H. DUSSAUCE. 12mo. $5 00

**DUPLAIS.—A COMPLETE TREATISE ON THE DISTILLATION AND MANUFACTURE OF ALCOHOLIC LIQUORS:** From the French of M. DUPLAIS. Translated and Edited by M. MCKENNIE, M D. Illustrated by numerous large plates and wood engravings of the best apparatus calculated for producing the finest products. In one vol. royal 8vo. $10 00

☞ This is a treatise of the highest scientific merit and of the greatest practical value, surpassing in these respects, as well as in the variety of its contents, any similar volume in the English language.

**DE GRAFF.—THE GEOMETRICAL STAIR-BUILDERS' GUIDE:** Being a Plain Practical System of Hand-Railing, embracing all its necessary Details, and Geometrically Illustrated by 22 Steel Engravings; together with the use of the most approved principles of Practical Geometry. By SIMON DE GRAFF, Architect. 4to. . . . . . . . . . . . . . $5 00

**DYER AND COLOR-MAKER'S COMPANION:** Containing upwards of two hundred Receipts for making Colors, on the most approved principles, for all the various styles and fabrics now in existence; with the Scouring Process, and plain Directions for Preparing, Washing-off, and Finishing the Goods. In one vol. 12mo. . . . . . . $1 25

**EASTON.—A PRACTICAL TREATISE ON STREET OR HORSE-POWER RAILWAYS:**

Their Location, Construction, and Management; with General Plans and Rules for their Organization and Operation; together with Examinations as to their Comparative Advantages over the Omnibus System, and Inquiries as to their Value for Investment; including Copies of Municipal Ordinances relating thereto. By ALEXANDER EASTON, C. E. Illustrated by 23 plates, 8vo., cloth . . . . . . . . $2 00

**FORSYTH.—BOOK OF DESIGNS FOR HEAD-STONES, MURAL, AND OTHER MONUMENTS:**

Containing 78 Elaborate and Exquisite Designs. By FORSYTH. 4to., cloth . . . . . . . . . . . . $5 00

*** This volume, for the beauty and variety of its designs, has never been surpassed by any publication of the kind, and should be in the hands of every marble-worker who does fine monumental work.

**FAIRBAIRN.—THE PRINCIPLES OF MECHANISM AND MACHINERY OF TRANSMISSION:**

Comprising the Principles of Mechanism, Wheels, and Pulleys, Strength and Proportions of Shafts, Couplings of Shafts, and Engaging and Disengaging Gear. By WILLIAM FAIRBAIRN, Esq., C. E., LL. D., F. R. S., F. G. S., Corresponding Member of the National Institute of France, and of the Royal Academy of Turin; Chevalier of the Legion of Honor, etc. etc. Beautifully illustrated by over 150 wood-cuts. In one volume 12mo. $2 50

**FAIRBAIRN.—PRIME-MOVERS:**

Comprising the Accumulation of Water-power; the Construction of Water-wheels and Turbines; the Properties of Steam; the Varieties of Steam-engines and Boilers and Wind-mills. By WILLIAM FAIRBAIRN, C. E., LL. D., F. R. S., F. G. S. Author of "Principles of Mechanism and the Machinery of Transmission." With Numerous Illustrations. In one volume. (In press.)

**GILBART.—A PRACTICAL TREATISE ON BANKING:**

By JAMES WILLIAM GILBART. To which is added: THE NATIONAL BANK ACT AS NOW IN FORCE. 8vo. . . . $4 50

**GESNER.—A PRACTICAL TREATISE ON COAL, PETROLEUM, AND OTHER DISTILLED OILS.**

By ABRAHAM GESNER, M. D., F. G. S. Second edition, revised and enlarged. By GEORGE WELTDEN GESNER, Consulting Chemist and Engineer. Illustrated. 8vo. . . . $3 50

**GOTHIC ALBUM FOR CABINET MAKERS:**
Comprising a Collection of Designs for Gothic Furniture. Illustrated by twenty-three large and beautifully engraved plates. Oblong . . . . . . . . . $3 00

**GRANT.—BEET-ROOT SUGAR AND CULTIVATION OF THE BEET:**
By E. B. GRANT. 12mo. . . . . . . $1 25

**GREGORY.—MATHEMATICS FOR PRACTICAL MEN:**
Adapted to the Pursuits of Surveyors, Architects, Mechanics, and Civil Engineers. By OLINTHUS GREGORY. 8vo., plates, cloth . . . . . . . . . . . $3 00

**GRISWOLD.—RAILROAD ENGINEER'S POCKET COMPANION.**
Comprising Rules for Calculating Deflection Distances and Angles, Tangential Distances and Angles, and all Necessary Tables for Engineers; also the art of Levelling from Preliminary Survey to the Construction of Railroads, intended Expressly for the Young Engineer, together with Numerous Valuable Rules and Examples. By W. GRISWOLD. 12mo., tucks. $1 75

**GUETTIER.—METALLIC ALLOYS:**
Being a Practical Guide to their Chemical and Physical Properties, their Preparation, Composition, and Uses. Translated from the French of A. GUETTIER, Engineer and Director of Founderies, author of "La Fouderie en France," etc. etc. By A. A. FESQUET, Chemist and Engineer. In one volume, 12mo. $3 00

**HATS AND FELTING:**
A Practical Treatise on their Manufacture. By a Practical Hatter. Illustrated by Drawings of Machinery, &c., 8vo. $1 25

**HAY.—THE INTERIOR DECORATOR:**
The Laws of Harmonious Coloring adapted to Interior Decorations: with a Practical Treatise on House-Painting. By D. R. HAY, House-Painter and Decorator. Illustrated by a Diagram of the Primary, Secondary, and Tertiary Colors. 12mo. $2 25

**HUGHES.—AMERICAN MILLER AND MILLWRIGHT'S ASSISTANT:**
By WM. CARTER HUGHES. A new edition. In one volume, 12mo. . . . . . . . . . $1 50

**HUNT.—THE PRACTICE OF PHOTOGRAPHY.**

By Robert Hunt, Vice-President of the Photographic Society, London. With numerous illustrations. 12mo., cloth . 75

**HURST.—A HAND-BOOK FOR ARCHITECTURAL SURVEYORS:**

Comprising Formulæ useful in Designing Builders' work, Table of Weights, of the materials used in Building, Memoranda connected with Builders' work, Mensuration, the Practice of Builders' Measurement, Contracts of Labor, Valuation of Property, Summary of the Practice in Dilapidation, etc. etc. By J. F. Hurst, C. E. 2d edition, pocket-book form, full bound $2 50

**JERVIS.—RAILWAY PROPERTY:**

A Treatise on the Construction and Management of Railways; designed to afford useful knowledge, in the popular style, to the holders of this class of property; as well as Railway Managers, Officers, and Agents. By John B. Jervis, late Chief Engineer of the Hudson River Railroad, Croton Aqueduct, &c. One vol. 12mo., cloth . . . . . $2 00

**JOHNSON.—A REPORT TO THE NAVY DEPARTMENT OF THE UNITED STATES ON AMERICAN COALS:**

Applicable to Steam Navigation and to other purposes. By Walter R. Johnson. With numerous illustrations. 607 pp. 8vo., . . . . . . $10 00

**JOHNSTON.—INSTRUCTIONS FOR THE ANALYSIS OF SOILS, LIMESTONES, AND MANURES.**

By J. W. F. Johnston. 12mo. . . . . . 35

**KEENE.—A HAND-BOOK OF PRACTICAL GAUGING,**

For the Use of Beginners, to which is added a Chapter on Distillation, describing the process in operation at the Custom House for ascertaining the strength of wines. By James B. Keene, of H. M. Customs. 8vo. . . . $1 25

**KENTISH.—A TREATISE ON A BOX OF INSTRUMENTS,**

And the Slide Rule; with the Theory of Trigonometry and Logarithms, including Practical Geometry, Surveying, Measuring of Timber, Cask and Malt Gauging, Heights, and Distances. By THOMAS KENTISH. In one volume. 12mo. . . $1 25

**KOBELL.—ERNI.—MINERALOGY SIMPLIFIED:**

A short method of Determining and Classifying Minerals, by means of simple Chemical Experiments in the Wet Way. Translated from the last German Edition of F. VON KOBELL, with an Introduction to Blowpipe Analysis and other additions. By HENRI ERNI, M. D., Chief Chemist, Department of Agriculture, author of "Coal Oil and Petroleum." In one volume. 12mo. . . . . . $2 50

**LANDRIN.—A TREATISE ON STEEL:**

Comprising its Theory, Metallurgy, Properties, Practical Working, and Use. By M. H. C. LANDRIN, Jr., Civil Engineer. Translated from the French, with Notes, by A. A. FESQUET, Chemist and Engineer. With an Appendix on the Bessemer and the Martin Processes for Manufacturing Steel, from the Report of ABRAM S. HEWITT, United States Commissioner to the Universal Exposition, Paris, 1867. 12mo. . . $3 00

**LARKIN.—THE PRACTICAL BRASS AND IRON FOUNDER'S GUIDE.**

A Concise Treatise on Brass Founding, Moulding, the Metals and their Alloys, etc.; to which are added Recent Improvements in the Manufacture of Iron, Steel by the Bessemer Process, etc. etc. By JAMES LARKIN, late Conductor of the Brass Foundry Department in Reany, Neafie & Co.'s Penn Works, Philadelphia. Fifth edition, revised, with extensive Additions. In one volume. 12mo. . . . . . $2 25

**LEAVITT.—FACTS ABOUT PEAT AS AN ARTICLE OF FUEL:**
With Remarks upon its Origin and Composition, the Localities in which it is found, the Methods of Preparation and Manufacture, and the various Uses to which it is applicable; together with many other matters of Practical and Scientific Interest. To which is added a chapter on the Utilization of Coal Dust with Peat for the Production of an Excellent Fuel at Moderate Cost, especially adapted for Steam Service. By H. T. LEAVITT. Third edition. 12mo. . . . . $1 75

**LEROUX.—A PRACTICAL TREATISE ON THE MANUFACTURE OF WORSTEDS AND CARDED YARNS:**
Translated from the French of CHARLES LEROUX, Mechanical Engineer, and Superintendent of a Spinning Mill. By Dr. H. PAINE, and A. A. FESQUET. Illustrated by 12 large plates. In one volume 8vo. . . . . . . . . . . $5 00

**LESLIE (MISS).—COMPLETE COOKERY:**
Directions for Cookery in its Various Branches. By MISS LESLIE. 60th edition. Thoroughly revised, with the addition of New Receipts. In 1 vol. 12mo., cloth . . $1 50

**LESLIE (MISS). LADIES' HOUSE BOOK:**
a Manual of Domestic Economy. 20th revised edition. 12mo., cloth . . . . . . . . . . . $1 25

**LESLIE (MISS).—TWO HUNDRED RECEIPTS IN FRENCH COOKERY.**
12mo. . . . . . . . . . . . 50

**LIEBER.—ASSAYER'S GUIDE:**
Or, Practical Directions to Assayers, Miners, and Smelters, for the Tests and Assays, by Heat and by Wet Processes, for the Ores of all the principal Metals, of Gold and Silver Coins and Alloys, and of Coal, etc. By OSCAR M. LIEBER. 12mo., cloth $1 25

**LOVE.—THE ART OF DYEING, CLEANING, SCOURING, AND FINISHING:**
On the most approved English and French methods; being Practical Instructions in Dyeing Silks, Woollens, and Cottons, Feathers, Chips, Straw, etc.; Scouring and Cleaning Bed and Window Curtains, Carpets, Rugs, etc.; French and English Cleaning, etc. By THOMAS LOVE. Second American Edition, to which are added General Instructions for the Use of Aniline Colors. 8vo. . . . . . . . . . . . 5 00

**MAIN AND BROWN.—QUESTIONS ON SUBJECTS CONNECTED WITH THE MARINE STEAM-ENGINE:**

And Examination Papers; with Hints for their Solution. By THOMAS J. MAIN, Professor of Mathematics, Royal Naval College, and THOMAS BROWN, Chief Engineer, R. N. 12mo., cloth $1 50

**MAIN AND BROWN.—THE INDICATOR AND DYNAMOMETER:**

With their Practical Applications to the Steam-Engine. By THOMAS J. MAIN, M. A. F. R., Ass't Prof. Royal Naval College, Portsmouth, and THOMAS BROWN, Assoc. Inst. C. E., Chief Engineer, R. N., attached to the R. N. College. Illustrated. From the Fourth London Edition. 8vo. . . . . $1 50

**MAIN AND BROWN.—THE MARINE STEAM-ENGINE.**

By THOMAS J. MAIN, F. R. Ass't S. Mathematical Professor at Royal Naval College, and THOMAS BROWN, Assoc. Inst. C. E. Chief Engineer, R. N. Attached to the Royal Naval College. Authors of "Questions Connected with the Marine Steam-Engine," and the "Indicator and Dynamometer." With numerous Illustrations. In one volume 8vo. . . . . . $5 00

**MARTIN.—SCREW-CUTTING TABLES, FOR THE USE OF MECHANICAL ENGINEERS:**

Showing the Proper Arrangement of Wheels for Cutting the Threads of Screws of any required Pitch; with a Table for Making the Universal Gas-Pipe Thread and Taps. By W. A. MARTIN, Engineer. 8vo. . . . . . . . . . 50

**MILES—A PLAIN TREATISE ON HORSE-SHOEING.**

With Illustrations. By WILLIAM MILES, author of "The Horse's Foot"

**MOLESWORTH.—POCKET-BOOK OF USEFUL FORMULÆ AND MEMORANDA FOR CIVIL AND MECHANICAL ENGINEERS.**

By GUILFORD L. MOLESWORTH, Member of the Institution of Civil Engineers, Chief Resident Engineer of the Ceylon Railway. Second American from the Tenth London Edition. In one volume, full bound in pocket-book form . . . . $2 00

**MOORE.—THE INVENTOR'S GUIDE:**

Patent Office and Patent Laws: or, a Guide to Inventors, and a Book of Reference for Judges, Lawyers, Magistrates, and others. By J G. MOORE. 12mo., cloth . . . . . . $1 25

**NAPIER.—A MANUAL OF ELECTRO-METALLURGY:**

Including the Application of the Art to Manufacturing Processes. By JAMES NAPIER. Fourth American, from the Fourth London edition, revised and enlarged. Illustrated by engravings. In one volume, 8vo. . . . . . . . . . . $2 00

**NAPIER.—A SYSTEM OF CHEMISTRY APPLIED TO DYEING:**
By JAMES NAPIER, F. C. S. A New and Thoroughly Revised Edition, completely brought up to the present state of the Science, including the Chemistry of Coal Tar Colors. By A. A. FESQUET, Chemist and Engineer. With an Appendix on Dyeing and Calico Printing, as shown at the Paris Universal Exposition of 1867, from the Reports of the International Jury, etc. Illustrated. In one volume 8vo., 400 pages . . . . . $5 00

**NEWBERY.—GLEANINGS FROM ORNAMENTAL ART OF EVERY STYLE;**
Drawn from Examples in the British, South Kensington, Indian, Crystal Palace, and other Museums, the Exhibitions of 1851 and 1862, and the best English and Foreign works. In a series of one hundred exquisitely drawn Plates, containing many hundred examples. By ROBERT NEWBERY. 4to. . . . . . $15 00

**NICHOLSON.—A MANUAL OF THE ART OF BOOK-BINDING:**
Containing full instructions in the different Branches of Forwarding, Gilding, and Finishing. Also, the Art of Marbling Book-edges and Paper. By JAMES B. NICHOLSON. Illustrated. 12mo. cloth . . . . . . . . . . $2 25

**NORRIS.—A HAND-BOOK FOR LOCOMOTIVE ENGINEERS AND MACHINISTS:**
Comprising the Proportions and Calculations for Constructing Locomotives; Manner of Setting Valves; Tables of Squares, Cubes, Areas, etc. etc. By SEPTIMUS NORRIS, Civil and Mechanical Engineer. New edition. Illustrated, 12mo., cloth $2 00

**NYSTROM.—ON TECHNOLOGICAL EDUCATION AND THE CONSTRUCTION OF SHIPS AND SCREW PROPELLERS:**
For Naval and Marine Engineers. By JOHN W. NYSTROM, late Acting Chief Engineer U. S. N. Second edition, revised with additional matter. Illustrated by seven engravings. 12mo. $2 50

**O'NEILL.—A DICTIONARY OF DYEING AND CALICO PRINTING:**
Containing a brief account of all the Substances and Processes in use in the Art of Dyeing and Printing Textile Fabrics: with Practical Receipts and Scientific Information. By CHARLES O'NEILL, Analytical Chemist; Fellow of the Chemical Society of London; Member of the Literary and Philosophical Society of Manchester; Author of "Chemistry of Calico Printing and Dyeing." To which is added An Essay on Coal Tar Colors and their Application to

Dyeing and Calico Printing. By A. A. FESQUET, Chemist and Engineer. With an Appendix on Dyeing and Calico Printing, as shown at the Exposition of 1867, from the Reports of the International Jury, etc. In one volume 8vo., 491 pages . . $6 00

**OSBORN.—THE METALLURGY OF IRON AND STEEL:**

Theoretical and Practical: In all its Branches; With Special Reference to American Materials and Processes. By H. S. OSBORN, LL. D., Professor of Mining and Metallurgy in Lafayette College, Easton, Pa. Illustrated by 230 Engravings on Wood, and 6 Folding Plates. 8vo., 972 pages . . . . . . $10 00

**OSBORN.—AMERICAN MINES AND MINING:**

Theoretically and Practically Considered. By Prof. H. S. OSBORN, Illustrated by numerous engravings. 8vo. (*In preparation.*)

**PAINTER, GILDER, AND VARNISHER'S COMPANION:**

Containing Rules and Regulations in everything relating to the Arts of Painting, Gilding, Varnishing, and Glass Staining, with numerous useful and valuable Receipts; Tests for the Detection of Adulterations in Oils and Colors, and a statement of the Diseases and Accidents to which Painters, Gilders, and Varnishers are particularly liable, with the simplest methods of Prevention and Remedy. With Directions for Graining, Marbling, Sign Writing, and Gilding on Glass. To which are added COMPLETE INSTRUCTIONS FOR COACH PAINTING AND VARNISHING. 12mo., cloth, $1 50

**PALLETT.—THE MILLER'S, MILLWRIGHT'S, AND ENGINEER'S GUIDE.**

By HENRY PALLETT. Illustrated. In one vol. 12mo. . $3 00

**PERKINS.—GAS AND VENTILATION.**

Practical Treatise on Gas and Ventilation. With Special Relation to Illuminating, Heating, and Cooking by Gas. Including Scientific Helps to Engineer-students and others. With illustrated Diagrams. By E. E. PERKINS. 12mo., cloth . . . $1 25

**PERKINS AND STOWE.—A NEW GUIDE TO THE SHEET-IRON AND BOILER PLATE ROLLER:**

Containing a Series of Tables showing the Weight of Slabs and Piles to Produce Boiler Plates, and of the Weight of Piles and the Sizes of Bars to Produce Sheet-iron; the Thickness of the Bar Gauge in Decimals; the Weight per foot, and the Thickness on the Bar or Wire Gauge of the fractional parts of an inch; the Weight per sheet, and the Thickness on the Wire Gauge of Sheet-iron of various dimensions to weigh 112 lbs. per bundle; and the conversion of Short Weight into Long Weight, and Long Weight into Short. Estimated and collected by G. H. PERKINS and J. G. STOWE . . . . . . . . . . . . $2 50

**PHILLIPS AND DARLINGTON.—RECORDS OF MINING AND METALLURGY:**

Or, Facts and Memoranda for the use of the Mine Agent and Smelter. By J. ARTHUR PHILLIPS, Mining Engineer, Graduate of the Imperial School of Mines, France, etc., and JOHN DARLINGTON. Illustrated by numerous engravings. In one vol. 12mo. . $2 00

**PRADAL, MALEPEYRE, AND DUSSAUCE.—A COMPLETE TREATISE ON PERFUMERY:**

Containing notices of the Raw Material used in the Art, and the Best Formulæ. According to the most approved Methods followed in France, England, and the United States. By M. P. PRADAL, Perfumer-Chemist, and M. F. MALEPEYRE. Translated from the French, with extensive additions, by Prof. H. DUSSAUCE. 8vo. $10

**PROTEAUX.—PRACTICAL GUIDE FOR THE MANUFACTURE OF PAPER AND BOARDS.**

By A. PROTEAUX, Civil Engineer, and Graduate of the School of Arts and Manufactures, Director of Thiers's Paper Mill, 'Puy-de-Dômé. With additions, by L. S. LE NORMAND. Translated from the French, with Notes, by HORATIO PAINE, A. B., M. D. To which is added a Chapter on the Manufacture of Paper from Wood in the United States, by HENRY T. BROWN, of the "American Artisan." Illustrated by six plates, containing Drawings of Raw Materials, Machinery, Plans of Paper-Mills, etc. etc. 8vo. $5 00

**REGNAULT.—ELEMENTS OF CHEMISTRY.**

By M. V. REGNAULT. Translated from the French by T. FORREST BENTON, M. D., and edited, with notes, by JAMES C. BOOTH, Melter and Refiner U. S. Mint, and WM. L. FABER, Metallurgist and Mining Engineer. Illustrated by nearly 700 wood engravings. Comprising nearly 1500 pages. In two vols. 8vo., cloth $10 00

**REID.—A PRACTICAL TREATISE ON THE MANUFACTURE OF PORTLAND CEMENT:**

By HENRY REID, C. E. To which is added a Translation of M. A. Lipowitz's Work, describing a new method adopted in Germany of Manufacturing that Cement. By W. F. REID. Illustrated by plates and wood engravings. 8vo. . . . . . . $7 00

**RIFFAULT, VERGNAUD, AND TOUSSAINT.—A PRACTICAL TREATISE ON THE MANUFACTURE OF COLORS FOR PAINTING:**

Containing the best Formulæ and the Processes the Newest and in most General Use. By MM. RIFFAULT, VERGNAUD, and TOUSSAINT. Revised and Edited by M. F. MALEPEYRE and Dr. EMIL WINCKLER. Illustrated by Engravings. In one vol. 8vo. (*In preparation.*) . . . . . .

**RIFFAULT, VERGNAUD, AND TOUSSAINT.—A PRACTICAL TREATISE ON THE MANUFACTURE OF VARNISHES:**

By MM. RIFFAULT, VERGNAUD, and TOUSSAINT. Revised and Edited by M. F. MALEPEYRE and Dr. EMIL WINCKLER. Illustrated. In one vol. 8vo. (*In preparation.*)

**SHUNK.—A PRACTICAL TREATISE ON RAILWAY CURVES AND LOCATION, FOR YOUNG ENGINEERS.**

By WM. F. SHUNK, Civil Engineer. 12mo., tucks . . $2 00

**SMEATON.—BUILDER'S POCKET COMPANION:**

Containing the Elements of Building, Surveying, and Architecture; with Practical Rules and Instructions connected with the subject. By A. C. SMEATON, Civil Engineer, etc. In one volume, 12mo. . . . . . . . . . . . . $1 50

**SMITH.—THE DYER'S INSTRUCTOR:**

Comprising Practical Instructions in the Art of Dyeing Silk, Cotton, Wool, and Worsted, and Woollen Goods: containing nearly 800 Receipts. To which is added a Treatise on the Art of Padding; and the Printing of Silk Warps, Skeins, and Handkerchiefs, and the various Mordants and Colors for the different styles of such work. By DAVID SMITH, Pattern Dyer, 12mo., cloth $3 00

**SMITH.—THE PRACTICAL DYER'S GUIDE:**

Comprising Practical Instructions in the Dyeing of Shot Cobourgs, Silk Striped Orleans, Colored Orleans from Black Warps, ditto from White Warps, Colored Cobourgs from White Warps, Merinos, Yarns, Woollen Cloths, etc. Containing nearly 300 Receipts, to most of which a Dyed Pattern is annexed. Also, a Treatise on the Art of Padding. By DAVID SMITH. In one vol. 8vo. $25 00

**SHAW.—CIVIL ARCHITECTURE:**

Being a Complete Theoretical and Practical System of Building, containing the Fundamental Principles of the Art. By EDWARD SHAW, Architect. To which is added a Treatise on Gothic Architecture, &c. By THOMAS W. SILLOWAY and GEORGE M. HARDING, Architects. The whole illustrated by 102 quarto plates finely engraved on copper. Eleventh Edition. 4to. Cloth. $10 00

**SLOAN.—AMERICAN HOUSES:**

A variety of Original Designs for Rural Buildings. Illustrated by 26 colored Engravings, with Descriptive References. By SAMUEL SLOAN, Architect, author of the "Model Architect," etc. etc. 8vo. $2 50

**SCHINZ.—RESEARCHES ON THE ACTION OF THE BLAST-FURNACE.**

By CHAS. SCHINZ. Seven plates. 12mo. . . . $4 25

**SMITH.—PARKS AND PLEASURE GROUNDS:**
Or, Practical Notes on Country Residences, Villas, Public Parks, and Gardens. By CHARLES H. J. SMITH, Landscape Gardener and Garden Architect, etc. etc. 12mo. . . . . $2 25

**STOKES.—CABINET-MAKER'S AND UPHOLSTERER'S COMPANION:**
Comprising the Rudiments and Principles of Cabinet-making and Upholstery, with Familiar Instructions, Illustrated by Examples for attaining a Proficiency in the Art of Drawing, as applicable to Cabinet-work; The Processes of Veneering, Inlaying, and Buhl-work; the Art of Dyeing and Staining Wood, Bone, Tortoise Shell, etc. Directions for Lackering, Japanning, and Varnishing; to make French Polish; to prepare the Best Glues, Cements, and Compositions, and a number of Receipts, particularly for workmen generally. By J. STOKES. In one vol. 12mo. With illustrations $1 25

**STRENGTH AND OTHER PROPERTIES OF METALS.**
Reports of Experiments on the Strength and other Properties of Metals for Cannon. With a Description of the Machines for Testing Metals, and of the Classification of Cannon in service. By Officers of the Ordnance Department U. S. Army. By authority of the Secretary of War. Illustrated by 25 large steel plates. In 1 vol. quarto . . . . . . . . . $10 00

**SULLIVAN.—PROTECTION TO NATIVE INDUSTRY.**
By Sir EDWARD SULLIVAN, Baronet. (1870.) 8vo. . $1 50

**TABLES SHOWING THE WEIGHT OF ROUND, SQUARE, AND FLAT BAR IRON, STEEL, ETC.**
By Measurement. Cloth . . . . . . . 63

**TAYLOR.—STATISTICS OF COAL:**
Including Mineral Bituminous Substances employed in Arts and Manufactures; with their Geographical, Geological, and Commercial Distribution and amount of Production and Consumption on the American Continent. With Incidental Statistics of the Iron Manufacture. By R. C. TAYLOR. Second edition, revised by S. S. HALDEMAN. Illustrated by five Maps and many wood engravings. 8vo., cloth . . . . . . . . . . $6 00

**TEMPLETON.—THE PRACTICAL EXAMINATOR ON STEAM AND THE STEAM-ENGINE:**
With Instructive References relative thereto, for the Use of Engineers, Students, and others. By WM. TEMPLETON, Engineer. 12mo. $1 25

**THOMAS.—THE MODERN PRACTICE OF PHOTOGRAPHY.**

By R. W. THOMAS, F. C. S. 8vo., cloth . . . . 75

**THOMSON.—FREIGHT CHARGES CALCULATOR.**

By ANDREW THOMSON, Freight Agent . . . . $1 25

**TURNING: SPECIMENS OF FANCY TURNING EXECUTED ON THE HAND OR FOOT LATHE:**

With Geometric, Oval, and Eccentric Chucks, and Elliptical Cutting Frame. By an Amateur. Illustrated by 30 exquisite Photographs. 4to. . . . . . . . . . . . $3 00

**TURNER'S (THE) COMPANION:**

Containing Instructions in Concentric, Elliptic, and Eccentric Turning; also various Plates of Chucks, Tools, and Instruments; and Directions for using the Eccentric Cutter, Drill, Vertical Cutter, and Circular Rest; with Patterns and Instructions for working them. A new edition in 1 vol. 12mo. $1 50

**URBIN—BRULL.—A PRACTICAL GUIDE FOR PUDDLING IRON AND STEEL.**

By ED. URBIN, Engineer of Arts and Manufactures. A Prize Essay read before the Association of Engineers, Graduate of the School of Mines, of Liege, Belgium, at the Meeting of 1865–6. To which is added a COMPARISON OF THE RESISTING PROPERTIES OF IRON AND STEEL. By A. BRULL. Translated from the French by A. A. FESQUET, Chemist and Engineer. In one volume, 8vo. $1 00

**VOGDES.—THE ARCHITECT'S AND BUILDER'S POCKET COMPANION AND PRICE BOOK.**

By F. W. VOGDES, Architect. Illustrated. Full bound in pocket-book form. . . . . . . . . . . . . $2 00

In book form, 18mo., muslin . . . . . . . 1 50

**WARN.—THE SHEET METAL WORKER'S INSTRUCTOR, FOR ZINC, SHEET-IRON, COPPER AND TIN PLATE WORKERS, &c.**

By REUBEN HENRY WARN, Practical Tin Plate Worker. Illustrated by 32 plates and 37 wood engravings. 8vo. . . $3 00

**WATSON.—A MANUAL OF THE HAND-LATHE.**

By EGBERT P. WATSON, Late of the "Scientific American," Author of "Modern Practice of American Machinists and Engineers," In one volume, 12mo. . . . . . . $1 50

**WATSON.—THE MODERN PRACTICE OF AMERICAN MACHINISTS AND ENGINEERS:**

Including the Construction, Application, and Use of Drills, Lathe Tools, Cutters for Boring Cylinders, and Hollow Work Generally, with the most Economical Speed of the same, the Results verified by Actual Practice at the Lathe, the Vice, and on the Floor. Together with Workshop management, Economy of Manufacture, the Steam-Engine, Boilers, Gears, Belting, etc. etc. By EGBERT P. WATSON, late of the "Scientific American." Illustrated by eighty-six engravings. 12mo. . . . . . . $2 50

**WATSON.—THE THEORY AND PRACTICE OF THE ART OF WEAVING BY HAND AND POWER:**

With Calculations and Tables for the use of those connected with the Trade. By JOHN WATSON, Manufacturer and Practical Machine Maker. Illustrated by large drawings of the best Power-Looms. 8vo. . . . . . . . . . . . . . $10 00

**WEATHERLY.—TREATISE ON THE ART OF BOILING SUGAR, CRYSTALLIZING, LOZENGE-MAKING, COMFITS, GUM GOODS,**

And other processes for Confectionery, &c. In which are explained, in an easy and familiar manner, the various Methods of Manufacturing every description of Raw and Refined Sugar Goods, as sold by Confectioners and others . . . . $2 00

**WILL.—TABLES FOR QUALITATIVE CHEMICAL ANALYSIS.**

By Prof. HEINRICH WILL, of Giessen, Germany. Seventh edition. Translated by CHARLES F. HIMES, Ph. D., Professor of Natural Science, Dickinson College, Carlisle, Pa. . . $1 25

**WILLIAMS.—ON HEAT AND STEAM:**

Embracing New Views of Vaporization, Condensation, and Expansion. By CHARLES WYE WILLIAMS, A. I. C. E. Illustrated. 8vo. $3 50

**WORSSAM.—ON MECHANICAL SAWS:**

From the Transactions of the Society of Engineers, 1867. By S. W. WORSSAM, Jr. Illustrated by 18 large folding plates. 8vo. $5 00

**WÖHLER.—A HAND-BOOK OF MINERAL ANALYSIS.**

By F. WÖHLER. Edited by H. B. NASON, Professor of Chemistry, Rensselaer Institute, Troy, N. Y. With numerous Illustrations. 12mo. . . . . . . . . . . . . . $3 00

www.ingramcontent.com/pod-product-compliance
Lightning Source LLC
LaVergne TN
LVHW011230110826
845150LV00006B/1591

* 9 7 8 1 4 2 5 5 1 5 1 4 0 *